SANITATION for Foodservice Workers

SANITATION for Foodservice Workers

TREVA M. RICHARDSON

SECOND EDITION

CAHNERS BOOKS INTERNATIONAL, INC.
221 Columbus Avenue
Boston, Massachusetts 02116
Affiliated with the Publishers of Institutions/Volume Feeding Magazine

Library of Congress Cataloging in Publication Data

Richardson, Treva M
Sanitation for foodservice workers.

1. Food service--Sanitation. 2. Food handling.
I. Title.
TX943.R5 1974 642 74-23606

Third Printing

ISBN 0-8436-0582-0

Printed in the United States of America

CONTENTS

1 INTRODUCTION TO FOODSERVICE SANITATION 1

2 BACTERIOLOGY 5

3 COMMUNICABLE DISEASES 16

4 FOOD-BORNE DISEASES 22

5 FOOD PROTECTION 37

6 PEST CONTROL 47

THE TEACHING OF DISHROOM PROCEDURES 55

7 MANUAL DISHWASHING 56

8 DISHWASHING COMPOUNDS 63

9 DETERGENT DISPENSING SYSTEMS 72

10 MECHANICAL DISHWASHING 82

11 SANITIZING 97

12 RESPONSIBILITIES FOR DISHROOM MAINTENANCE 110

13 PERSONAL HYGIENE 118

14 OUTLINES FOR FOLLOW-UP MEETINGS 128

ACKNOWLEDGEMENTS

I would like to express my appreciation for source material and background information provided by the following for this book:

Personnel of the Center for Disease Control, Atlanta, Georgia.

Manufacturers of equipment used by the foodservice industry.

Manufacturers of dishwashing compounds and other related products used in cleaning in foodservice establishments.

Manufacturers of pesticides used by the foodservice industry.

Personnel of Health Departments, teachers, school lunch supervisors, foodservice managers and other industry workers who offered many knowledgeable suggestions for this revision.

I would also like to acknowledge my debt to Wade Richardson Nicodemus for his invaluable assistance in the preparation and writing of this revised edition of *Sanitation for Food Service Workers.*

Treva M. Richardson

SANITATION
for Foodservice Workers

1 Introduction to Foodservice Sanitation

Sanitation for Foodservice Workers is a manual designed to educate foodservice personnel about sanitation and also to provide practical training in the regulations and procedures necessary to maintain a successful sanitation program in a foodservice establishment. Such education and training must include more than the passing on of information. Certain rules, procedures and correct personal habits must become such an integral part of daily work routines as to be practically automatic in performance. This very desirable situation is not easy to accomplish. Health workers have been trying various approaches for years.

By the time of the United States' entry into World War II, a few dedicated public health workers, in widely scattered areas, had already begun to experiment with sanitation training programs for foodservice personnel.

One of those most interested in personnel training was Dr. W. H. Haskell of the United States Public Health Service. A born teacher, Dr. Haskell possessed a unique ability to explain the basic facts and necessary procedures in a manner easily understood by personnel. His listeners were convinced of the value of the instruction and, in most cases, could realize the necessity for following the suggested routines.

Dr. Haskell used slides to illustrate the basic initial training course. His discussion-type method of presentation varied according to the class. A slide could be held on the screen while the points concerning it were covered simply or it could be held longer for more detailed, in-depth discussion. He based the length of discussion on whether the class members were newly employed personnel, experienced workers, supervisors, managers or interested citizen groups. His remarks were tailored to the needs of the class.

Whenever possible, appropriate posters and flip charts were displayed in the classroom. Leaflets or booklets containing condensed reminders of the basic facts were given to students during the final session.

In these early days of sanitation training, all such aids were in short supply. The need for their further development was realized when the training of foodservice workers was intensified during World War II. Health agencies and industry suppliers began the increased production of film strips, moving pictures and printed educational matter which continues at an accelerated pace today.

Today's courses on Sanitation for Foodservice Workers can be set up to include slide films, moving pictures and other visual aids while still having as a foundation the six main sessions (subdivided as indicated on p. 2) using a minimum of 176 slides to illustrate the basic sanitation information.

This revised edition of *Sanitation for Foodservice Workers* has been expanded to provide newer and more detailed information in certain areas and to respond to today's increasing interest in foodservice sanitation.

Many of today's foodservice employees read about foodservice methods and happenings in newspapers and other publications or hear about them on TV and radio. They are often made curious by what they read and hear and want to know how it applies in their own activities or the activities of those around them. They may ask questions pertaining to diseases involved in foodservice work, the latest method of detergent dispensing or similar subjects. This revised edition has been expanded to provide the instructing supervisor with ready answers to such questions.

THE COURSE MATERIAL

The material is easily adaptable to the varied abilities and attitudes of any type of class. Sufficient coverage is provided to permit in-depth management study; or, in initial employee courses where time is a factor, the instructor may stress those facts most in need of emphasis in his particular situation, leaving the remainder for subsequent follow-up meetings. The text has been prepared with follow-up in mind.

Follow-up is as essential to personnel training as is the proper presentation of the course itself. No matter how well the class material is taught, employees will require constant assistance through channels of periodic post-course discussions, visual aids such as posters, etc. and quiet on-duty reminders of any sanitation lapses.

The course has been presented in several ways. The following set-up of six sessions has proved effective for on-premise presentation.

SESSION I

Bacteriology for Foodservice Personnel
Communicable Disease

SESSION II

Food-Borne Diseases—Food Poisoning and Food Infection

SESSION III

Food Protection
Pest Control

SESSION IV

Manual Dishwashing
Dishwashing Compounds—
Detergent Dispensers

SESSION V

Mechanical Dishwashing
Sanitizing
Responsibilities for Maintenance

SESSION VI

Personal Hygiene
Question and Answer Review

Many commercial establishments hold a class once a day for six days, then repeat the course a second six days for those who remain on the job during the first sessions. In this way, there is no interruption of service.

Presenting sanitation instruction to small groups is a method being used by an increasing number of foodservice operations. Perhaps this is because foodservice personnel fall naturally into various small groupings making it easier for matters pertinent to the work of each group to be discussed in separate sessions with the supervisor. For example, in any establishment of some size there are dining room personnel (divided into bus people, waiters, etc. according to the individual establishment's set-up); the salad department; pastry department; chef's station (also subdivided according to duties); dishroom personnel; storeroom employees; clean-up and maintenance, etc.

In a small establishment, the employees work together as a team and everyone knows a good deal about the other fellow's job. It is easy to learn—and accept—the sanitation requirements for each station and for the operation as a whole. In the larger establishment, it is more difficult to learn the specifics of each sanitary routine required at each station.

There are several ways to use the small group concept in foodservice training. If management needs to present the material in a two-week time period, taking part of the personnel from their jobs the first week, then repeating the course a second week for the balance of the staff, the following plan is suggested.

Cover Bacteriology and Communicable Disease with the group as a whole, devoting one class period to each subject. Bacteriology is the foundation of the entire course. Both topics need concentrated study and uninterrupted attention from the personnel, with plenty of time for slide showing, a question and answer period, and general discussion.

With the exception of Personal Hygiene, the other subjects could be covered partly in class, partly throughout the establishment. Allowing one hour for each topic, spend 15 minutes on the basics, 30 minutes in small groups, 15 minutes as a whole group in a concluding wrap-up session.

Following this method, after a firm

grounding in the basic material, the class breaks up into groups. Each small group will visit a different station each day of the course, thus the entire establishment will have been covered by the end of the training. Group members could be selected by the instructor, or could consist of pastry workers, salad department workers, storeroom personnel, etc. Several supervisors or knowledgeable employees should be available to head the groups and answer questions.

The purpose of these visits to different departments is *not* to find fault with another group's work. Rather it is to better understand the sanitation rules pertaining to each department by viewing departments at first hand from a new perspective. Also, the inspecting group could note the sanitation danger points in each area with a more objective eye than that used by the department's regular workers. Notes should be taken and short reports made by the various leaders of the small groups when the class assembles in the meeting room for the closing remarks.

If the time element is not important in management's planning for sanitation training, several variations in presentation are possible.

Management may prefer to go through the entire course first with all personnel in one group, then divide them into small groups for the following week's meeting.

Or, after a general discussion, the class could be divided into small groups and continue the topic with a group leader, making notes, asking questions, etc. The session would close with comments from each leader. Department visits could follow later.

After a general session, each department's personnel could also meet alone to further discuss pertinent points.

Or at a special meeting, one small group could lead and demonstrate sanitary procedures. For example, the whole class could meet in the dishroom while the crew demonstrates just what is required to produce tableware that meets cleanliness standards.

Small group discussions increase interest by making for easier communication between management and personnel. Most employees feel more relaxed about expressing opinions, offering suggestions, asking questions in a small group than they would in a meeting of all personnel. The group leader notes all suggestions, etc. and makes a report when the groups re-form into the general session.

Educators have always advocated sanitation training for management—training in more depth involving a study of regulations, equipment, foods, preparation, storage, serving, etc. In several large cities, where it is impossible for sanitarians to make frequent visits to each establishment and where personnel turnover may be frequent, health departments are conducting management courses with notable success. Such courses are invaluable in operations where self-inspection is carried on.

Here again, it is suggested that management become thoroughly familiar with the text, slides and method used in *Sanitation for Foodservice Workers* before the material is presented to the employees. It is becoming more and more the rule for the manager and supervisory personnel to take the course first—in more depth, including presentation of the teaching methods to be used by prospective supervisor-instructors. Local health department sanitarians will be of great assistance with this undertaking.

Notes to Instructors:

As with any type of instruction, the teacher is largely responsible for the success of the presentation. Sanitation training is actually a form of adult education. This means that adult students are not going to blindly accept or adopt the necessary sanitation rules and procedures without understanding why they should be accepted and incorporated into daily routines.

The instructor, therefore, must know his material thoroughly but should not parrot words from a memorized script. He needs to be an out-going individual who enjoys talking to people. He should develop the knack of establishing an informal, relaxed classroom atmosphere while at the same time gaining and maintaining the student's interest and close attention. Above all, his presentation should be sincere. The class must be convinced he believes in the sanitary procedures he is teaching.

Introductory remarks to the class should employ an approach designed to arouse student/personnel interest in the material even before the sessions begin. A brief discussion along the following lines will aid in achieving the desired result:

1. The classes will teach employees not only how to provide protected foods and sanitized equipment and utensils for patrons but will also establish essential sanitation practices to protect the personnel.

2. The conscientious application of sanitation routines will raise the professional status of personnel by increasing their efficiency, and enhancing their appearance; their exhibition of assured "know-how" and deftness during proper preparation and serving of food will have a most favorable impression on both patrons and management. Increased self-confidence and pride in their work will be the inevitable result.

3. The information learned will be found useful in the home, where it can be applied to protect the health of families and friends.

Dr. Haskell, pioneer in foodservice sanitation training, always read the slides aloud with the class, dwelling on any points which were not clear. He wanted this developed into an ad lib presentation which would vary with each lecture. Some of Dr. Haskell's accompanying comments for some of the slides are reproduced on these pages. No attempt has been made to present an exact wording inasmuch as the comments were deliberately varied to avoid a stereotyped approach. Dr. Haskell after reading material contained in the slide, asked questions on the lecture material presented prior to the slide's showing, and utilized such slides as an oral quiz. This evoked interest and audience participation.

Editor's Note

Treva M. Richardson is one of hundreds trained in food sanitation by the late Dr. W. H. Haskell. During World War II she worked with him in the Chicago office of the U. S. Public Health Service, leaving, in 1947, to become Superintendent of Sanitation for the Pixley & Ehlers chain of restaurants. In 1954, a move to Florida brought about an association with the Morrison Cafeteria Company which was terminated by an illness in 1963.

Mrs. Richardson now devotes herself to the writing of food sanitation articles, many of which have appeared in Institutions/VF Magazine. She also acts as a consultant in foodservice personnel training methods and problems.

Her collaboration with Dr. Haskell in the preparation of personnel training material and her possession of his original unpublished notes enable her to offer an accurate transcription of his training methods.

The 176 2- by 2-in. slides illustrating the subjects covered in "Sanitation for Food Service Workers" are available for rental or purchase. For slide information please write Mrs. Treva M. Richardson, 220 S. Logan Street, Mason City, Illinois 62664.

---

2 Bacteriology

We are beginning our training in sanitation with a brief discussion of Bacteriology. And the first thing we must understand is *why* it is necessary and right for us to start with a subject which only a few of you may have studied previously—a subject which, you may be thinking, is not very practical to include in such a down-to-earth topic as sanitation.

When most of us think of sanitation we think of cleanliness . . . spotless floors, polished furniture, shining dishes, gleaming silver-ware, immaculate refrigerators, clean uniforms and clean people wearing those uniforms. All of these could be grouped under the general heading of good housekeeping, and being good housekeepers.

All these are important to us and to the people who eat the food we prepare and serve. It is more comfortable and easier to work in pleasant, clean surroundings. Food becomes more enjoyable in a clean atmosphere. More important, the people who eat that food assume such a clean place is sanitary. To the customers, "sanitary" means not only that the establishment is clean, but that the food and drink are wholesome and safe, with no danger of their causing customer illness.

These assumptions are partially correct. Good housekeeping is one of the most essential factors to the success of a sanitation program. But good housekeeping alone will not ensure a sanitary state of affairs throughout the establishment, neither will it prevent food poisonings nor the transmission of certain other diseases through food and drink.

In the final analysis, the fate of a sanitation program depends, to a large extent, on the thinking, the beliefs and the actions of the operation's personnel. And before personnel can think and act correctly they must understand what is really meant by sanitation and why it is their responsibility to practice sanitary procedures. Workers should understand that sanitation is needed to protect both their own health and that of the people who trust the operation's management enough to eat the food and drink prepared on the premises.

So we come back to Bacteriology which —in the admittedly over-simplified definition considered sufficient for our purposes—we shall refer to as the study of bacteria, or microbes or germs. You will hear all three terms mentioned during our training course on foodservice sanitation.

The simple fact is that were it not for the existence of germs, there would be no need for sanitation. Once we learn enough about germs to understand why and how some of the dangerous ones are involved in foodservice and why certain rules for coping with them are necessary, we will have come a long way in our understanding of sanitation.

On the other hand, until we do come to appreciate the importance of bacteria in the proper preparing and serving of food and drink, and until we do have a better understanding of how bacteria live, multiply and get about, we shall not have sanitation either in business or in the home.

Since we are not approaching the study of Bacteriology in the routine manner, using microscopes, watching live germs through them and so forth, we are going to be rather unconventional about this. We will make you familiar with bacteria and their habits by comparing them with people—real people like yourselves.

In spite of the fact that bacteria are so tiny you cannot see them with the naked eye, which in itself makes understanding that much more difficult, you will soon discover that people and germs resemble each other in several ways. There are also a few very significant ways in which they are not like people.

Let us first consider the resemblances. Bacteria, like human beings, must be actually alive to do either good or evil. So it is the possibility of encountering living bacteria, especially the dangerous ones, and how to prevent such situations that concern us here.

Germs Vs. Humans

(1) Resemblances

Both:

1. Are alive.	**5. React to chemicals.**
2. Eat food.	**6. Colonize.**
3. Give off waste products.	**7. Do damage to persons or things**
4. React to temperature.	**8. Aid humanity**

Human beings must have food in order to live, and so must bacteria. Both germs and people thrive on many of the same foods. Milk, milk products, eggs, meat, poultry, fish, shellfish. . .these are among the foods supporting the growth of disease-causing bacteria (pathogens). Still others in this category include cream pies, custards; potato, ham, poultry and fish salads; cheese, hollandaise sauce, poultry dressings.

These foods are among those considered "potentially hazardous"—that is, they can become dangerous to us because they support the rapid growth of bacteria which cause illness in human beings. It would seem that any food relished by human beings will also prove suitable for the growth and development of some form of bacterial life. This is why a working knowledge of how germs live and grow is a "must" for foodservice personnel.

Both people and germs enjoy moist foods. The germ's life, in fact, depends largely upon moisture, inasmuch as it obtains its nourishment through the very thin membrane which surrounds it. This eating procedure, called osmosis, is not complicated or difficult to understand if you will just visualize water soaking through the skin of a grape. The food that the bacteria require is dissolved in the water and thus passes into the bacteria.

People and germs alike give off waste products from their bodies. Certain people and certain germs throw off wastes which may cause illness. These wastes are chemicals and are called toxins. There are two kinds of toxins: exotoxins and endotoxins. Exotoxins are poisons usually secreted (given off) by living bacteria. They are very potent (strong) poisons. Among the diseases caused by exotoxic bacteria are the botulism and staphylococcus food poisonings, tetanus, diphtheria and scarlet fever. Some exotoxins, such as botulism toxin, can be destroyed by 15 minutes boiling but many others require much higher temperatures for destruction. The problem in food sanitation—a very real and serious problem—is that cooking or reheating may kill the toxin-producing bacteria but leave the toxin undamaged to cause a later food poisoning.

Endotoxins are poisons that are usually released after the death of the bacteria. Among the diseases caused by these bacteria are dysentery, typhoid fever, undulant fever and salmonella food poisoning. These toxins are very difficult to destroy, even at high temperatures. So they, too, pose a problem in the foodservice industry.

The best way to protect ourselves and others, then, from these "criminal" members of the bacterial society is to take firm action to either destroy them or prevent their dangerous build-up in food or in people. Most of these bacteria, once taken into the body, if still alive, will continue to

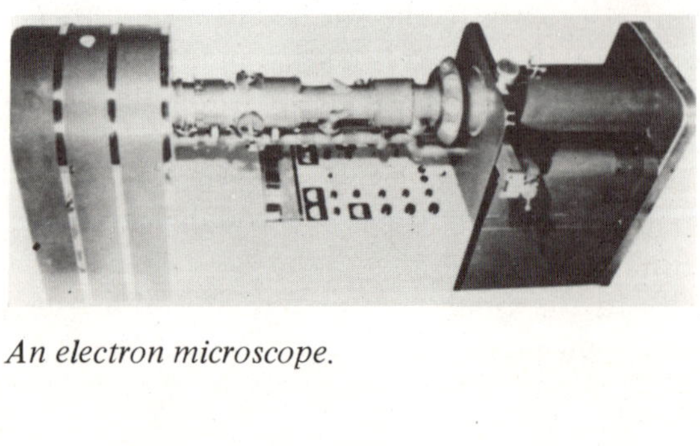

An electron microscope.

grow, multiply and produce toxins.* Abnormal conditions in food may also be caused by the waste products of germs, an important fact to be remembered.

What Is Proper Temperature?

There is a definite resemblance to the way people and germs are affected by various temperatures. Heat will kill members of both groups, provided the temperature is sufficiently high. However, there are some people—and some germs—who exist very comfortably in warm surroundings. Thus, we have heat resistant germs and heat resistant people.

Cold will retard the multiplication of bacteria and herein lies the value of refrigeration. When we use cold (refrigeration) to prevent the growth of bacteria, we must make sure that the temperature is cold enough to accomplish its purpose. Halfway measures will only defeat our purpose. There are some types of germs able to withstand lower temperatures than others.

Everyone knows that certain chemicals such as cyanide and arsenic can destroy human life. There are also chemicals which destroy germs. In the food industry, there are definite limitations to the choice of chemicals for this purpose. We must consider toxicity, poisoning, flavors and odors. Here again, when chemicals are used to destroy bacteria, the dilution must be strong enough to insure effectiveness.

Germs Will Colonize

We are all Americans, yet the families of many citizens came to America originally from other countries. Today, it is not unusual to see people of the same national and cultural backgrounds living near each other in the same neighborhoods. Bacteria also tend to live together in groups—or, as we say, colonize.

Sometimes when people of one group move into a colony of persons of a different group, trouble and violence occur. These same situations arise in colonies of bacteria.

We know that all types of people make up the human family. We have only to read the newspapers, listen to radio or watch television to learn of heroic actions, great achievements and many good deeds. But just as often we are told of many crimes—robberies, kidnappings, hijackings, muggings.

Mankind tries to protect itself against the destroyers and killers of society by giving them prison sentences or removing them permanently from contact with their fellow men. To control the bacterial killers and destroyers, to prevent them from harming human beings or food, we practice the rules of sanitation.

Let us continue this comparison even further. In the world of people, a breakdown in law enforcement will result in the release of some criminals or the escape from punishment of those who should have been kept away for life from the remainder of society. These individuals are then free to prey on others.

When dealing with germs, if the rules of sanitation are not conscientiously enforced, if sanitary procedures are not followed routinely, dangerous bacteria are left free to cause illness and then trouble results. An effective sanitation program is one which is enforced, actively and vigorously, on a daily basis.

The vast majority of people are not, of course, killers, robbers or property destroyers. Instead, they labor constructively to make useful products or carry on businesses important to our way of life. Many bacteria also work to bring about beneficial results. Butter, cheese, buttermilk, many crops are the result of helpful bacteria.

As a general rule, bacteria take their family names from the shapes they have. Many bacteria are rather round or oval in shape. These make up the "cocci" family (singular, coccus).

BACTERIA MAY BE CLASSIFIED BY SHAPE

Round in shape	—called *Cocci*
Rod shaped	—called *Bacilli*
Spiral shaped	—called *Spirilla*

Viruses—seen only under special conditions

MOLDS—YEASTS—PROTOZOA

---

*Note: Examples of both exotoxic and endotoxic bacteria have been listed above to give us a little more information about toxins and how they are formed. *Please note that not all toxin-forming bacteria are the causes of food poisonings.* Of the bacteria named above, only three are responsible for food poisonings: Staphylococci, Salmonella and the bacillus responsible for Botulism.

Others, shaped like tiny rods not unlike cigarettes, are the "bacilli" (singular, bacillus). The "spirilla" (singular, spirillum) make up the third group.

During these classes you will hear and read about a number of organisms with much longer or more complex names than coccus, bacillus or spirillum.

Sometimes you will recognize a familiar family name as part of another word or term. Example: staphlo*coccus*, strepto*coccus, bacillus* anthracis, etc. Just as often, the entire word may be unfamiliar to you—as, for example, clostridium perfringens or salmonella.

Do not try to remember all these words. For our purposes, it is important only that you realize these organisms are alive, that some of them are disease causing, and that, of these, a number may be transmitted through food and/or drink.

You might try to remember the shape and name of coccus, bacillus and spirilla. The terms *added* to these basic names are used to describe them more fully. These complicated words will be memorized by those working in the field of medical science.

You might try to remember the three general family names given to bacteria. Aside from these, the other identifying names are listed only to show how the smaller units of the larger family groups are subdivided so that each branch of the family can be known immediately. And even these branches can—and are—subdivided over and over again so that medical science can give an exact name for the bacterium responsible for a specific disease.

Differences Between Humans and Germs

As a contrast, let us now discuss a few of the ways in which bacteria do *not* resemble human beings. These differences are extremely important because they include some of the factors essential to the efficient application of sanitation.

SIZE—Bacteria vary in size and appearance just as humans. Germs, however, cannot be seen with the naked eye. The average bacterium is 1/25,000th of an inch in size. Consideration of size is very important. It is impossible for many people to realize they

HUMANS AND GERMS	
Differences	
Size	**Germs have no arms—legs—wings**
Reproduction	
Method of Eating	**Germs have no racial prejudices or politics**

are dealing with germs, simply because they are invisible. Things must be seen to be impressive.

The fact that bacteria cannot be seen is responsible for most of our sanitation failures. These failures will continue until such time as a more workable understanding of bacteria and a greater appreciation of what germs are capable of doing is fully realized.

As a comparison, 25,000 bacteria occupy 1 inch; 25,000 average people, shoul-

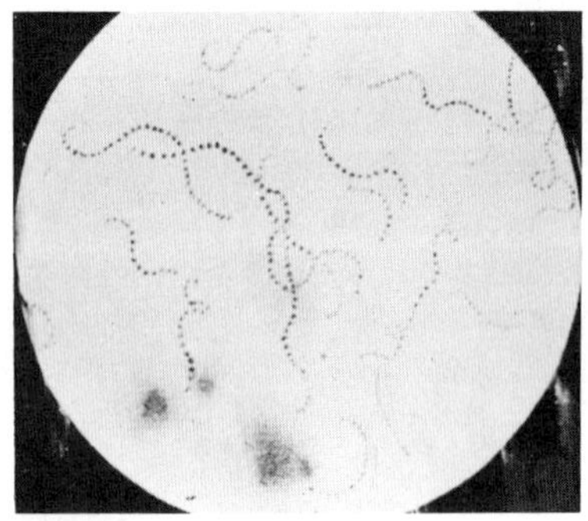

A streptococcus.

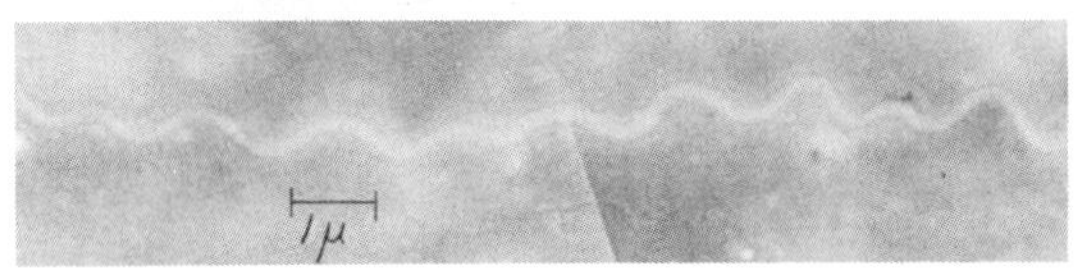

Treponema pallidum (syphilis)

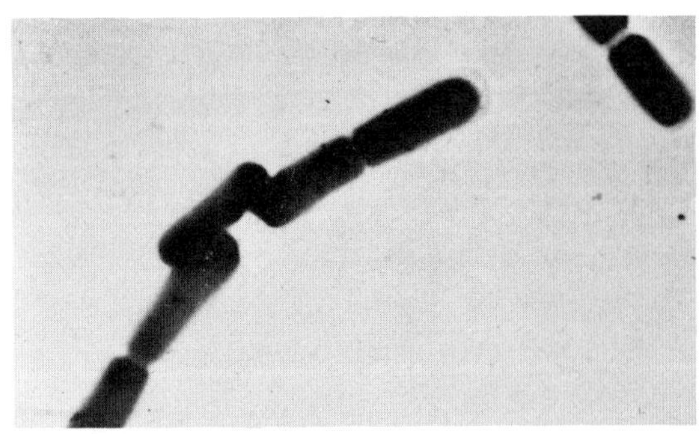

A botulinis organism. This will be shown to you again and discussed at length in your lecture on food poisoning and food intoxication. In this case, it is not the germ itself but the toxin (poison) it produces which causes illness.

THE REPRODUCTION RATE OF BACTERIA

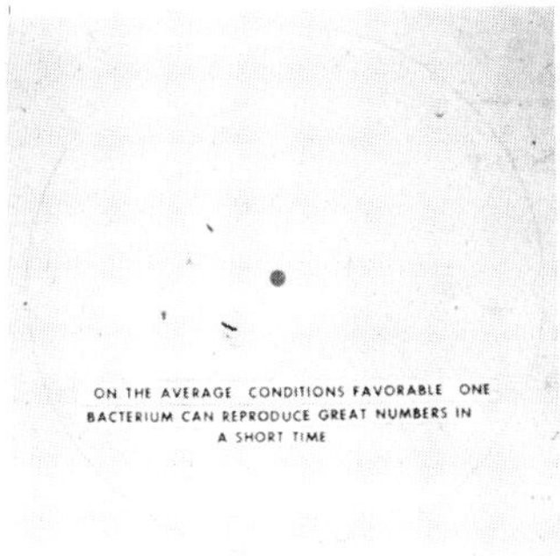

Reproduction rate of bacteria is well illustrated here. First, the individual bacterium . . . .

. . . after 20 min., two from the original one . . . .

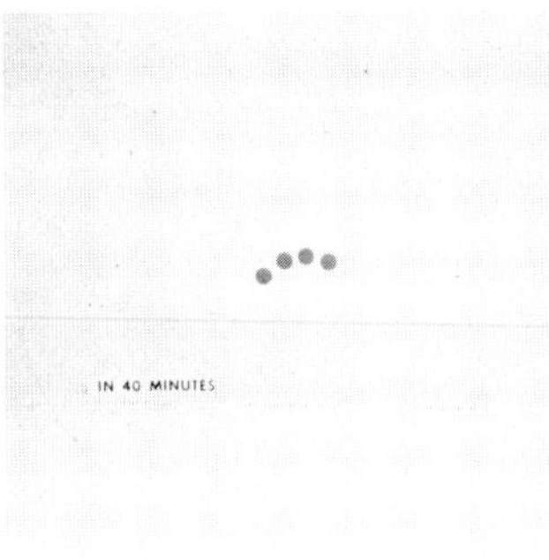

. . . after 40 min., four individual germs develop . .

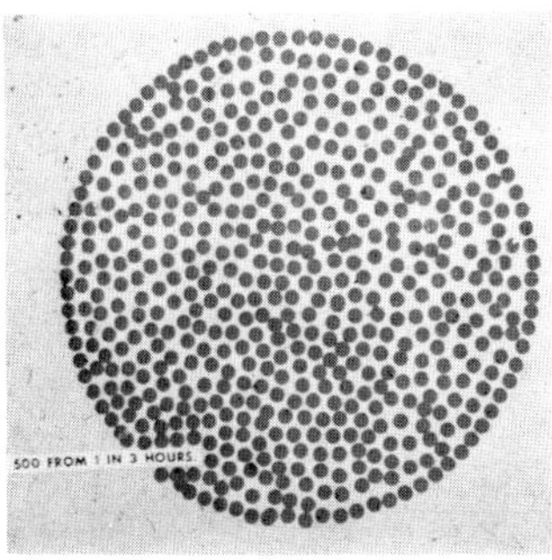

. . . after three hours, 500 organisms from that first one.

der to shoulder in a straight line would extend for a distance of over 10 miles.

REPRODUCTION–By simply dividing in half, bacteria reproduce themselves. The speed with which this division occurs varies with different types of germs. On the average, each germ, under favorable conditions, will divide and become two germs every 20 minutes.

If people multiplied as rapidly as bacteria, a couple with two babies, leaving the babies at home and spending three hours at a restaurant or show, would find 1000 babies when they returned home. If they had been away for five hours, they would return to find a city of 33,000 population.

Do you understand, from this rather ridiculous comparison, the very serious importance of preventing the multiplication of bacteria in food?

LOCOMOTION–Germs have no means of locomotion by which they can travel from place to place. They are not provided with arms, legs or wings. Germs must be transported from one place to another by some living thing or inanimate object. And this is the reason for sanitation.

Sanitation is composed of rules designed to prevent the transfer of germs. As we have mentioned previously, if germs did not exist, we would not need sanitation.

Germs are all around you–on your body, your hands, in your hair, on your faces and on your clothes. Some of them are harmless, some are harmful and some helpful. The main point to remember is that even if you cannot see them, you are going to have to deal with them.

Germs are the greatest hitchhikers in the world. They get about on contaminated objects, on people's hands, in various other ways connected with humans. Insects and rodents, dust and dirt, etc., also provide germs with the transportation they must have to get about. Hence the necessity for cleanliness, proper methods of handling utensils and food, as well as safeguarding water supplies, sewage disposal, insect and rodent control, etc.

Before we leave Bacteriology, we will cover briefly some of the other forms of cell life with which you should be familiar.

Sneezing into a culture dish containing a nutrient agar, in order to collect the bacteria in the droplets, Bacteria from droplets shown in tube below left.

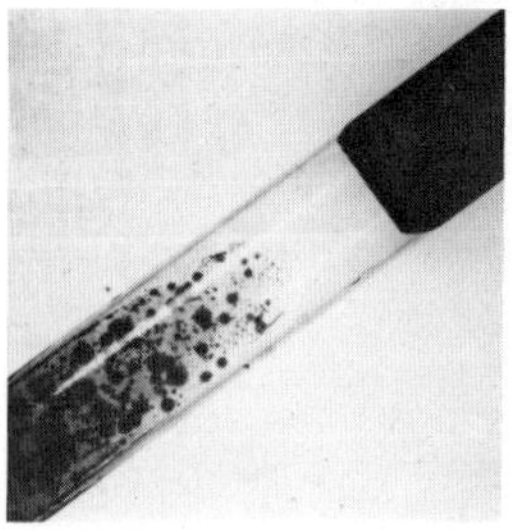

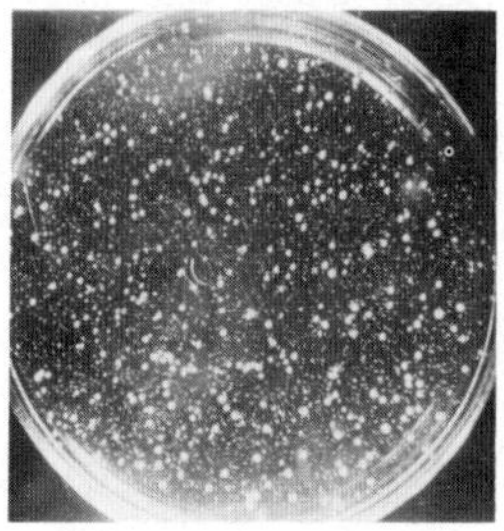

Colonies of streptococcus pyogenes growing on special media (right). Bacteria have food preferences just as people do. In the laboratory, knowing what types of food certain bacteria prefer, such a media is prepared to stimulate the growth of the germ. Germs grow in the form of colonies and these colonies may contain large numbers of bacteria.

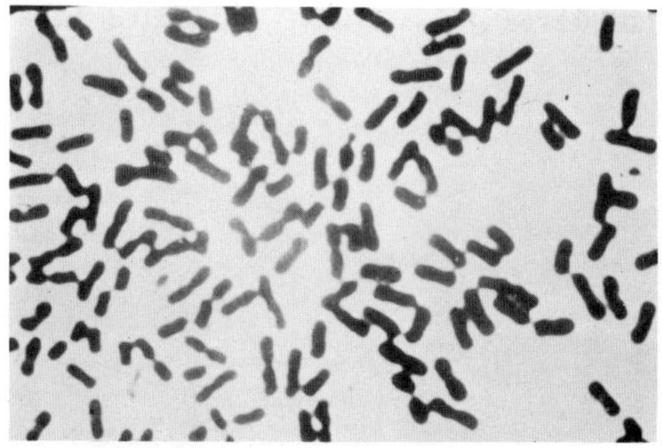

Germs ready to divide squeeze together in the middle, then break apart. Note good example top center, find others ready to separate.

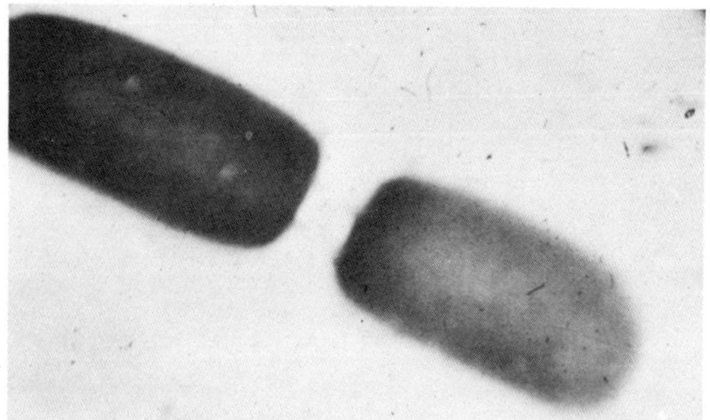

Escherichia coli, one of the coliforms, about ready to break apart. The only thing holding the two halves together is a very thin membrane that does not show in this illustration.

Molds

Molds belong to a form of life called "fungi" (singular, fungus). Fungi are plants that are not able to make their own food as trees, grass, flowers and most other plants can. A mold is made up of many cells, larger than those of bacteria. But, like most bacteria, the mold's food must be in a dissolved form.

Molds resemble true plants, since they develop numerous branches resembling threads or cobwebs. When these branches become extensive enough, the mold is easily visible as a powdered growth, rather like cotton.

Different types of molds have different colors—white, black, pink, green, blue. They may grow on sour milk, bread, cottage cheese, vegetables, fruits and other foods. Wood products, paper, etc. are also susceptible to mold growth.

Molds require air and moisture for growth. They like dampness and darkness. Temperature requirements vary with the type of mold. Some molds thrive in cold places. They have been found growing at 23°F.—nine degrees below the freezing point. These molds occur in refrigerators.

Most molds multiply by forming reproductive cells called spores. These spores are very light, can be found in large numbers in the soil (particularly dust) and can be carried by air currents, animals, insects and people.

As with bacteria, some types of molds are very helpful to mankind, others cause problems and disease. Antibiotics are manufactured from certain molds. Some of these medicines are well known to us—penicillin, terramycin, aureomycin, tetracycline, chloromycetin and many others.

Certain molds, grown purposely on cheeses such as Roquefort and Camembert, give such a special, different flavor to these cheeses they have become gourmet items.

This is not the case with most moldy foods. The unpleasant odor and musty flavor of moldy foods make them unpopular with most people. However, they are not believed to cause actual illness if eaten.

On the problem side, certain mold-like growths are responsible for mildews and such common skin infections as ringworm, athlete's foot and barber's itch.

Molds and mold spores can be destroyed by the same methods used to eliminate bacteria.

Protozoa

Protozoans are microscopic one-celled animals and are considered to be the oldest living type of animal life. Because they are so very small and defenseless, many of them serve as a food source for larger animals, which in turn, are eaten by larger animals, which in turn may be eaten by man. In other words, many protozoa are at the bottom of different food chains.

You have seen that bacteria, which are also one-celled, are grouped according to their shape: coccus, bacillus, spirillum. Protozoa are grouped according to what kind of structure moves them. If people were grouped in this way, there would be one type only, since we all use legs as our structure by which to move.

There are, however, three different kinds of structures used for movement in the protozoa. One type, called the pseudopods (pronounce it "soodo-pods") simply pushes out its cell membrane into a long leg and then flows into it. In this way it moves slowly from one place to another. The different Amoeba members belong to this group. A second group of protozoa move by means of short hairs which project from their cells. These hairs are called cilia and protozoans bearing them are called ciliates. The Paramecia are members of this group.

A third group possesses longer and fewer hairs called flagella and are referred to as the flagellates. The Euglena are widespread members of this group. A fourth group of protozoa are called the sporozoa because all the members produce spores. One member of this group causes malaria.

There are several other human diseases caused by protozoans. Among these are African sleeping sickness, dumdum fever, and amoebic dysentery. This latter disease is marked by abdominal pains, diarrhea, headache and fever. It is caused by a member of the Amoeba group called Entamoeba histolytica.

Viruses

Viruses have long been one of the more interesting and strange forms of life and are probably one of the most primitive life types. In some ways they are like living matter but in other ways they resemble nonliving matter.

Viruses are too small to be seen with even the most highpowered light microscope. They can be seen only with an electron microscope (p. 6). Viruses are not in the form of cells like bacteria, nor are they composed of cells like man. They are clumps of the same kind of chemical that make up chromosomes, the tiny structures of heredity. This chemical is enclosed in a protein coat.

Viruses do not take in food or oxygen like living things, but can reproduce with the aid of the cells which they attack.

Different viruses cause a wide range of diseases. These include measles, mumps, rabies, colds, polio, influenza, infectious hepatitis, and a variety of intestinal upsets that cause fever, vomiting, diarrhea, and abdominal pains. Several of these diseases may be transmitted through foods. Polio, infectious hepatitis and some intestinal infections have been transmitted in this manner.

Yeasts

Yeasts are another class of fungi. Like bacteria, they are one-celled organisms but are larger than the bacterial cells. Each cell absorbs food and moisture and reproduces. Yeasts multiply by forming buds. The buds enlarge until they are almost the same size as the mother cell when they are pinched off from the original cell and become independent cells.

COMPARATIVE SIZES OF VIRUSES

	Molecular weight x 10^6 (Particle weight x6 06 x 10^{17})	Diam or length x width in mμ
Red blood cells*	173,000,000	7500
Bacillus prodigiosus*	173,000	750
Rickettsia*	11,100	300
Psittacosis*	8,500	275
Vaccinia*	4,300	
Myxoma*	4,300	225
Canary pox*	4,300	
Pleuro-pneumonia organism*	1,400	
Pseudo rabies	1,400	150
Ectromelia	1,400	
Herpes simplex	1,400	
Rabies fixe	800	125
Borna disease	800	
Influenza	400	100
Vesicular stomatitis	400	
Staphylococcus bacteriophage†	300	90
Fowl plague	300	
C_{16} bacteriophage	173	75
Chicken tumor I*	142	70
Tobacco mosaic*	35	280 x 15
Cucumber mosaics 3 and 4*	35	
Gene (Muller's est. of max. size)*	33	125 x 20
Latent mosaic of potato*	26	430 x 9.8
Rabbit papilloma (Shope)*	25	40
Equine encephalitis	23	38
Megatherium bacteriophage	23	
Rift valley fever	11	30
Tomato bushy stunt*	8	26
Hemocyanin molecule (Busycon)*	6.7	22
Yellow fever	4.3	22
Tobacco ring spot*	3.4	19
Louping ill	2.8	19
Hemocyanin molecule (Octopus)*	2.8	20
Alfalfa mosaic*	2.1	16.5
Poliomyelitis	0.7	12
Staphylococcus bacteriophage†	0.4	10
Foot-and-mouth disease	0.4	
Hemoglobin molecule (Horse)*	0.069	15 x 3
Egg albumin molecule*	0.040	9 x 3

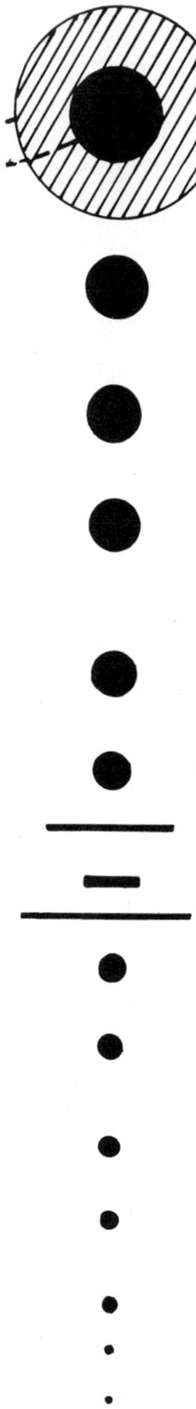

The various types of yeast will grow and multiply wherever air, moisture, suitable food and warm temperatures are combined. Under these conditions a yeast cell can grow to full size and reproduce in less than an hour.

Sugar is an essential yeast food, and yeasts have the ability to change starch into sugar. This ability is made use of, though in different ways, in the wine and baking industries.

We are familiar with some of the uses of yeast, such as in the making of bread. They are also important in the making of

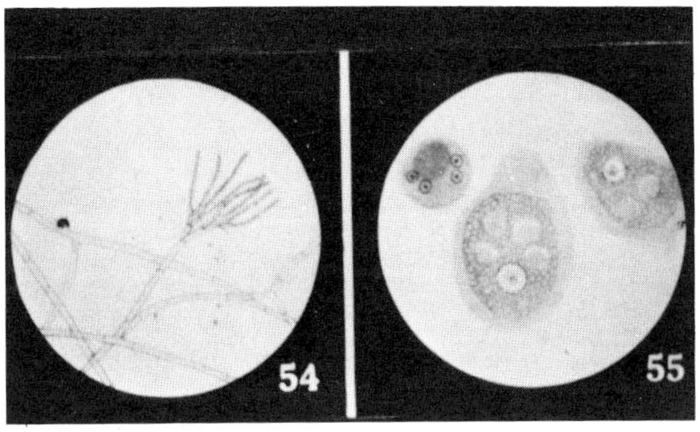

The budding of yeast

wine and other alcoholic beverages, as well as in various industrial processes. Among other uses, they are a vitamin B_{12} supplement.

Since yeast cells are very light, they may be air-borne or carried on the clothing or skin of man. Sometimes yeast cells fall into food and begin their breaking down process of the starches and sugars in the food. Food so contaminated will spoil.

Yeasts are not classified among the organisms considered dangerous to man, though a few types do cause some human skin infections.

The fact that the various types of organisms described above can—and do cause illness in human beings points up two important truths:

No. 1. Suitable conditions for the growth and multiplication of the organisms must exist. Otherwise they could not become numerous enough to do harm, either in themselves or in the production of toxins. Therefore, correct sanitation procedures in the storage, preparation, and serving of food, and in the washing and sanitizing of tableware and equipment are all "musts" in the protection of foods.

No. 2. Foodservice workers must be in good health on the job. They must carefully observe the rules of personal hygiene in order to protect themselves and others, both at work and at home.

Two methods are available to prevent the multiplication of germs once they enter food. They are described in a foodservice slogan and we give it to you here for the first time: *Keep the food hot or keep the food cold—and always keep it protected.*

As we go further and further along in this food sanitation training course, you will realize that each of the steps discussed has as its purpose the prevention of the transfer of germs from one place to another and from person to person.

Notes to Instructors

Foodservice personnel and students have shown keen interest in a simple experiment-demonstration which can be presented effectively near the close of the Bacteriology discussion, and finished during the concluding Personal Hygiene session.

The purpose of the experiment is to demonstrate that bacteria are alive; that they are everywhere, on the skin, clothing, utensils, articles of furniture, etc. We attempt to prove these statements by helping bacteria to grow in sufficient numbers to become visible to the human eye.

Materials needed for the experiment are: (1) four or five (or more) Petri dishes containing prepared agar solution, (a food on which bacteria will grow); (2) a package of sterile swabs and (3) small tweezers.

The required materials can usually be obtained from the local health department, a laboratory in the area (hospital or private), or from a high school or college science department. Health department personnel in many areas are very helpful in showing foodservice managers and others inexperienced in the use of these materials, how to handle them correctly.

The method followed is to remove bacteria from a selected object, transfer them to an agar solution, place the cover on the Petri dish, set the dish in a warm place where the temperature remains suitable for bacterial growth, and leave the dishes undisturbed until the final session at the end of the week.

The extent of the experiment will depend on the number of available Petri dishes of agar solution. It is interesting to have one dish contain a human hair, another a fingerprint and to let an employee or student cough into a third.

Ask the class for suggestions about where to transfer the bacteria, using some of the "tools of the trade." They might mention the can opener, a serving spoon, a clean table top, etc. Swab the suggested surface carefully, rub the swab over the surface of the agar solution, place cover on the Petri dish and tag each dish with the name of the object used.

As previously stated, the dishes should be left in a warm temperature, undisturbed, for several days. Then have the dishes brought to class, let the members study each one and discuss what they see.

If the experiment has been done correctly, colonies of bacteria, clearly visible to the naked eye, will be seen growing on the agar plates. This offers impressive proof of the presence of bacteria.

BACTERIOLOGY QUIZ

1. Name 3 resemblances between human beings and bacteria. ______________________
____
____

2. Name 3 differences between human beings and bacteria. ______________________
____
____

3. Fill in the following statements:
   Round shaped bacteria are called ____________
   Rod shaped bacteria are called ____________
   Spiral shaped bacteria are called ____________

4. Bacteria are good travellers; How do they get about? Check correct statement.
   (a) They fly ( )
   (b) They walk ( )
   (c) They hitchhike ( )

5. Besides bacteria, name three other forms of life which cause disease. ____________
____
____

6. Put a check after correct statements.
   (a) Bacteria can be seen with the naked eye. ( )
   (b) The average size bacterium is 1/25,000 of an inch in size. ( )
   (c) Bacteria lay eggs. ( )
   (d) Bacteria multiply by dividing in half. ( )

7. Good housekeeping alone will prevent disease outbreaks in a food operation.
   Check Yes ( ) No ( )

8. Put a check after correct statement.
   Germs eat by:
   (a) Chewing ( )
   (b) Drinking ( )
   (c) Absorbing food through cell wall. ( )

9. Some forms of life such as bacteria, give off chemicals called ____________________ which are poisonous to man.

10. Why are some bacteria dangerous? ________________________

3 Communicable Diseases

Some of the harmful bacteria we described in the chapter on Bacteriology are responsible for a group of illnesses classed as communicable diseases. Many of the most commonly known diseases fall into this category.

1. Why are these diseases called communicable?"

The establishing of communication between people or nations has become one of the popular topics for discussion in our time. An idea may be conveyed (carried) from one person to another; or a new policy or method of government may be passed along and shared by various groups of people or different nations.

In the case of certain diseases the bacteria which cause a specific illness are transferred or passed along from one person to another person This, too, is communication, so the diseases included in this group are called communicable. Unfortunately, in this case, harmful, disease-causing bacteria are passed along from one person to another.

2. Disease germs may travel from person to person in direct or indirect ways. When we learn (a) how this can be done, and (b) how to prevent germ transferrence by either the direct or indirect method, we will know how to protect our own health and that of others by following rules designed to prevent the germs of a communicable disease from being passed along from a sick individual to a well person.

In other words, we will practice sanitation.

An old saying states "If it were not for the existence of harmful bacteria, there would be no need to practice sanitation." This is worth remembering. It can help foodservice personnel realize that sanitation means more than achieving surface cleanliness. It means the carrying on of work routines in certain ways and the development of personal habits which will protect the worker's health and assist in preventing the spread of communicable (catching) forms of illness.

3. Foodservice workers have a special interest as well as a real involvement in studying communicable diseases. About 40 percent of the communicable diseases which physicians must officially report to health departments are associated with the business of providing food and/or drink to the public.

How necessary it is, then, for foodservice employees to understand how they can help prevent the spread of a communicable disease in the establishment where they are employed.

Knowing what to do is not enough. Learn to understand why certain rules have been drawn up. Then decide to follow those rules and carry out the sanitary routines as instructed.

4. Spreading Communicable Diseases

Following are listed the general ways in which the germs of communicable diseases may be spread:

(a) **Discharges from the respiratory tract.**

(i) The respiratory tract includes our speaking and breathing apparatus—mouth, nose, throat, lungs.

(ii) Discharges from the nose are often the first visible signs of what may become a serious illness. Measles, mumps, influenza, pneumonia, the common cold, meningitis, etc. are all infections involving discharges from the nose and throat. These discharges are usually heavily loaded with germs or viruses of the disease in question.

(iii) Sneezes and coughs may be blocked with the hands, a cloth handkerchief or disposable tissue. Let us consider what happens in each case.

First, the hands have become contaminated with germ laden droplets and must be thoroughly washed before handling food or utensils.

Second, handkerchiefs, once used, are contaminated. Hands in contact with soiled handkerchiefs become, without question, a source of infection for anything they touch. The temptation to use the same handkerchief to cover a cough or sneeze, blow or wipe the nose is practically irresistible. Hands used in this process must be washed before handling food, dishes or utensils.

(iv) Disposable tissues should be used once and discarded. Then wash the hands.

(b) **The intestinal tract.**

(i) In some communicable diseases the germs are present in the digestive tract and leave the body in intestinal discharges. These infections, often referred to as filth-borne diseases, include among others, typhoid fever, para-typhoid fever and the various types of dysentery.

(ii) Careless personal habits may result in the infection of food. This is the reason for the sign displayed in employee restrooms: "Wash hands before returning to work."

(c) **Insects and Rodents.**

Because of the importance of this subject, a full chapter has been devoted to it. "Pest Control" is its title, (pp. 47 to 53).

(d) **Infected persons**

(i) An infected person is one actually suffering from a communicable disease.

(ii) In addition to spreading the infection as described in sections (a) and (b), the person infected with a communicable disease goes through an incubation period. The incubation period, roughly, is from the time the bacteria first enter the body until the first symptoms of the disease appear. The time period varies with different diseases, plus the vitality of the germ and the resistance of the person may add or subtract from the incubation time. Examples: The common cold has an incubation period of from 12 to 48 hours; measles about 10 days.

COMMUNICABLE DISEASES

SOURCE	*PER CENT*
DIRECT OR INDIRECT CONTACT	43.3
CARRIERS	20.0
MILK-FOOD-WATER	30.0
INSECTS	26.5
RODENTS	20.0
MORE THAN ONE SOURCE	30.0

One major objective of this instruction is to teach foodservice personnel how to protect themselves against communicable disease brought in by customers.

THIS IS IMPORTANT TO YOU

(iii) The important thing to remember about an incubation period is: During this time, before the first symptom appears, the infected person may be giving off the germs of the disease without knowing that he has it. This is one more reason for foodservice personnel to observe the rules of food sanitation *at all times.*

(e) **Carriers**

A carrier is a person who, without apparent symptoms of communicable disease, carries the disease germs in the body and spreads them to other people. This spreading may occur directly from person to person or indirectly by way of utensils or food. Health ordinances regulating food establishments prohibit the employment of known carriers in foodservice work in areas or positions where there is any likelihood of transmission of disease to patrons or to fellow-workers, either through direct contact or through the contamination of food or food-contact surfaces with disease-causing organisms.

5. Germs may be spread *directly* from person to person or *indirectly* by way of utensils or food.

DIRECT CONTACT: Suppose someone is impolite or thoughtless enough to cough or sneeze into your face with no attempt made

A sizable part of the drinking rim of a cup, glass, mug, etc. actually enters the mouth. Hence the necessity for careful sanitization of dishes, cups, glasses and eating utensils. Bacteria left on these articles from previous use must be destroyed.

to block the droplet explosion with hands or tissues. Aside from being annoyed, you have every right to be concerned. Remember, the germ-laden droplets may travel as far as 12 feet. The chances are excellent that you will breathe in many of these droplets and thus introduce the germs of an infection into your own system.

This is an example of direct, person to person contact. In a home where there are children this can be the main reason for the rapid spread of a communicable disease throughout the family.

Direct contact, however, is not the way in which communicable diseases are most often passed along in a foodservice operation.

INDIRECT CONTACT: In this method of germ transference, harmful bacteria travel from an infected person to a well person by riding along on contaminated objects such as food, dishes, eating utensils, etc.

How do the germs reach these objects and contaminate them by their presence? And how may well persons be made ill when this occurs?

(a) The food or utensils may have been coughed or sneezed upon by an employee. This is not too likely but it must be considered.

(b) Fingers soiled by germ-laden discharges handle utensils and food Some of the germs remain on the food, dishes, cups, glasses, spoons, or other items.

(c) Insects and rodents contaminate food and equipment.

(d) The contaminated eating utensils are handled by well people. Spoons, forks, cups, glasses contact the mouth and the germs are thus transferred into the body of the well person.

(e) Food containing disease germs is eaten by a well person who may then become ill.

Hands

The hands of a foodservice worker contact many things during a work day. A large number of such contacts are involved with job procedures. Many others, unfortunately, include personal habits of long standing which are definitely undesirable in a food operation.

To test this statement, try, for just 15 minutes to keep a record of what a fellow foodservice worker does with his hands—what he touches, and what he picks up, etc. during the course of ordinary work activities.

You will undoubtedly see some of the

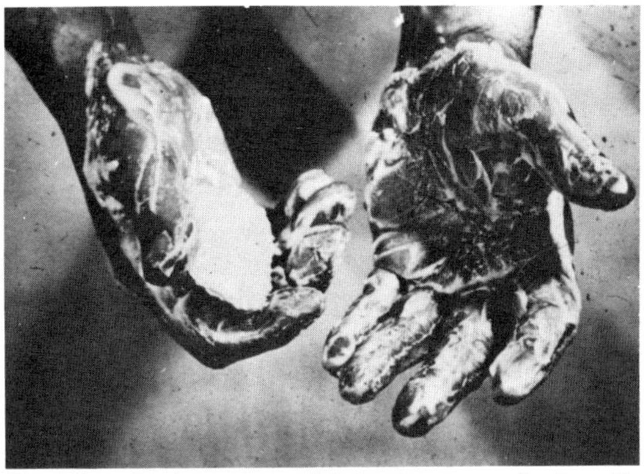

Even hands that seem visibly clean may be heavily contaminated with germs. But hands will never be worn out by washing. Good soap, warm water, plus plenty of friction (rubbing) are necessary in hand washing.

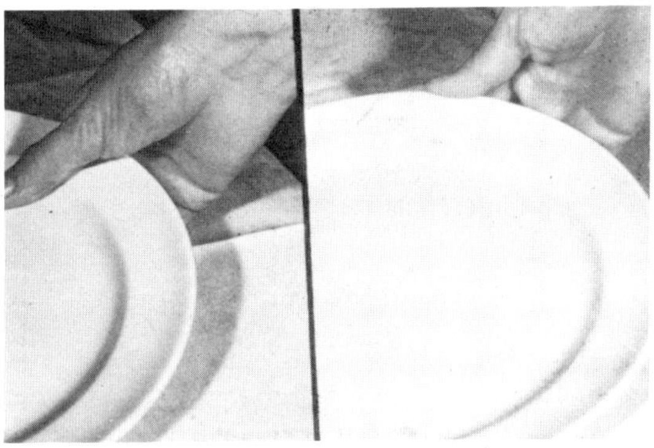

Thumb contact with food surface of plate; wrong at left, correct on right.

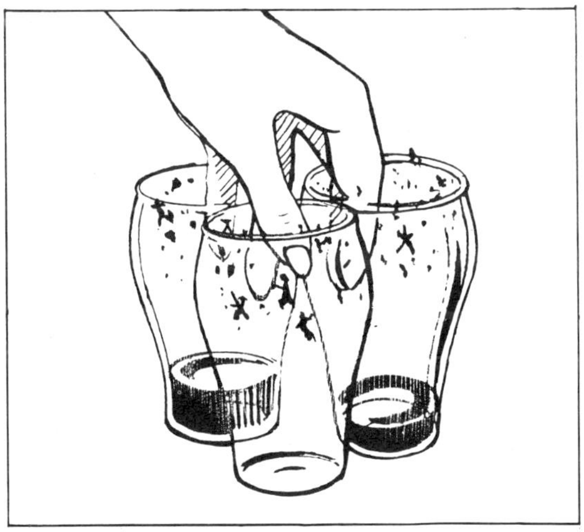

following, much of it done unconsciously.

(i) Fingers put in the mouth.

(ii) Face or head scratched.

(iii) Eyes or ears rubbed.

(iv) Nose rubbed, perhaps picked at.

(v) Face wiped with towel intended for counter wiping.

(vi) Sticking fingers into food for tasting purposes.

(vii) Sampling food with spoon, then putting spoon back into food.

And there are many more, including smoking on the job. The point of the habits listed above is this: If an infection (pimple, sore, cut, scratch, itch, boil, etc.) is present in any area of the body touched by fingers which next handle food or utensils, we have a perfect example of indirect transmission of germs.

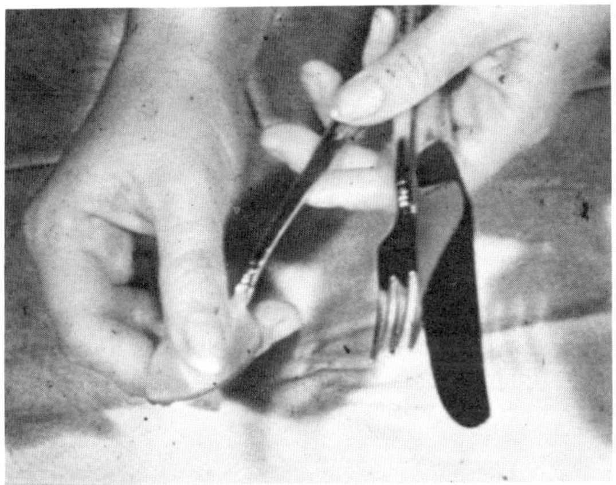

Worker should never put fingers on business end of utensils.

SUMMARY

A brief summary of important facts in this chapter includes:

1. Foodservice personnel must realize they may be dealing with communicable disease during the course of daily routines.
2. Such diseases may be passed from person to person by direct or indirect means.
3. Discharges from the nose, throat, eyes, ears must be kept off utensils and out of food.
4. Intestinal discharges must be kept away from equipment and food.
5. A person may be giving off bacteria of a communicable disease before any symptoms are evident.
6. Disease carriers should be considered as infected persons.
7. Utensils, equipment, foods, etc., all should be kept protected from germ-laden droplets due to coughing, sneezing or talking.
8. Hands provide one of the most common methods for the transmission of bacteria from place to place. Keep fingers out of, or off, the business parts of utensils, equipment, food.
9. Wash hands thoroughly after each restroom visit, following each handling of customer-soiled utensils, after coughing and sneezing or smoking.
10. Keep fingers out of the mouth.

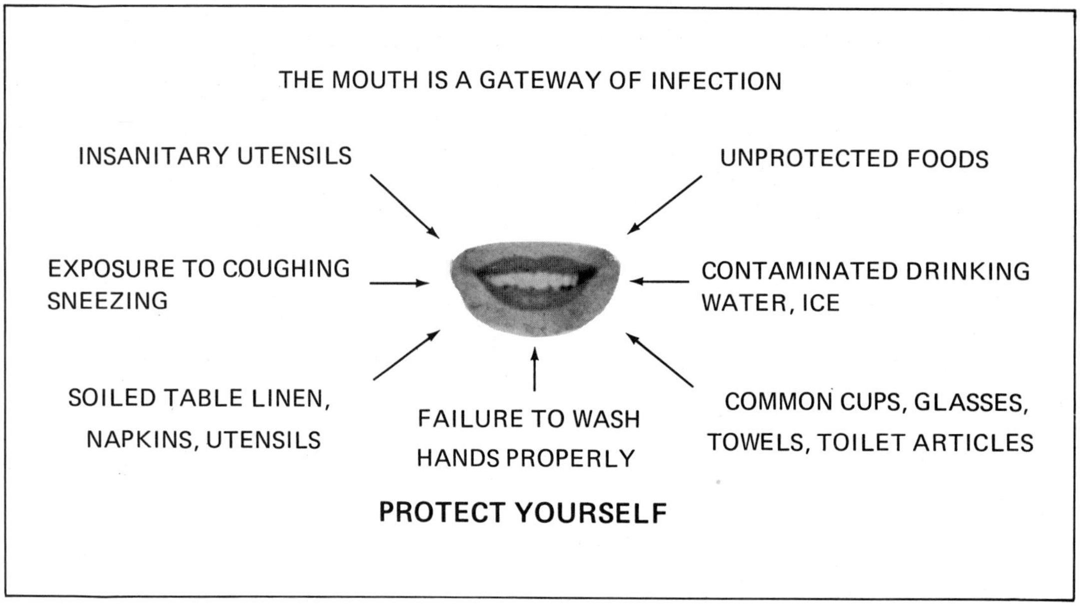

COMMUNICABLE DISEASE QUIZ

How much do you remember?

1. Name three of the ways in which germs may travel into the human body.
   (1) ______________ (2) ______________ (3) ______________

2. Put a check after the correct statement.
   Indirect contact means:
   (a) Coughing into someone's face. ( )
   (b) Coughing into food. ( )

3. True or False (check True or False):

	True	False
(a) Hands that look clean must be clean.	( )	( )
(b) It is OK to smoke if the ashes don't fall into the food.	( )	( )
(c) Fingers must be kept out of the mouth.	( )	( )

4. Diseases passed along from one person to another are called ______________.

5. Fill in the correct answers on the lines below.
   How may coughs or sneezes be blocked? Name three ways.
   1. ______________
   2. ______________
   3. ______________

6. Check correct statements.
   Incubation time means
   (a) A person has just recovered from a communicable disease. ( )
   (b) A person is catching a communicable disease. ( )
   (c) A person is sick with a communicable disease. ( )
   (d) A person is catching a communicable disease but does not feel ill. ( )

7. A person who has the germs of a communicable disease in his body but is not ill from the disease is called a ______________.

8. What is one of the most common methods for the transmission of bacteria from place to place? ______________

9. Name 3 times when hands need to be thoroughly washed.
   1. ______________
   2. ______________
   3. ______________

10. Give another name for communicable.
    ____

Note: The majority of reference material for this chapter was provided by the Center for Disease Control, Community Services Training Section, Public Health Service Bureau of Disease Prevention and Environmental Control, Atlanta, Ga.

4 Food-Borne Diseases

DEFINITIONS

1. "Food-borne disease" is a term applied to all types of illnesses caused by an organism, substance, or material of any kind which is present in food and which gains entrance to the body when such food is eaten.

2. Food Poisoning Intoxication is an illness caused from eating food which contains a poisonous substance. Such a substance could be a type of poison such as botulism toxin, produced by certain kinds of bacteria; a chemical or metallic poison, such as lead or arsenic; a poisonous substance present in the food because of natural causes, such as that found in poisonous mushrooms; or a poisonous substance found in the food due to physical causes such as radioactive fallout.

Many people, including foodservice personnel term all food-borne illnesses as food poisonings. This is not correct.

3. Food Infection is an illness caused by eating food or drink contaminated by certain specific organisms such as disease-causing bacteria. viruses, etc. When these organisms, such as Salmonella choloraesius, grow and multiply inside the body, with resulting action of the body's own chemical defenses, the symptoms peculiar to the particular infection appear.

FOOD POISONING ILLNESSES

1. Bacterial.

While certain types of bacteria, under suitable conditions, are growing and multiplying in food, toxins or poisons are produced. It is these toxins that cause illness, not the bacteria themselves. Hence, such food poisoning illnesses are properly termed Food Intoxications. The following are included in the Food Intoxication group:

(a) STAPHYLOCOCCUS FOOD POISONING—This is by far the most common of all food poisonings.

The staphylococcus organism is widespread. One authority states that approximately 50 percent of the general adult population have these germs present in their nasal passages and that 15 percent to 20 percent who carry the organisms in their nostrils also have them on the hands.

Staphylococci may be found in the throats of individuals, and in great abundance in the post-nasal discharges of persons recovering from colds. On the skin, staphylococci are found in acne, lesions, pimples, boils, carbuncles, infected cuts, infected burns, etc.

It is easy to understand how foodservice personnel could transfer these organisms to food. Droplets from the nose and throat, sneezed or coughed into the air, could contaminate air, clothing, skin, handkerchiefs or disposable tissues.

Hands using soiled handkerchiefs or tissues, hands touching the nose, or face, picking pimples or contacting other already contaminated skin areas become themselves, heavily soiled with these invisible germs. These same hands, working in food preparation areas, have many opportunities to transfer the contamination to utensils, equipment and any food they touch.

Foodservice personnel must realize the necessity for following sanitary procedures in food preparation, practicing correct personal hygiene and hands habits, and never permitting themselves to use even one careless, slap-dash food preparation method. The danger of contaminating food is too great, the opportunities too many, the ways too easy.

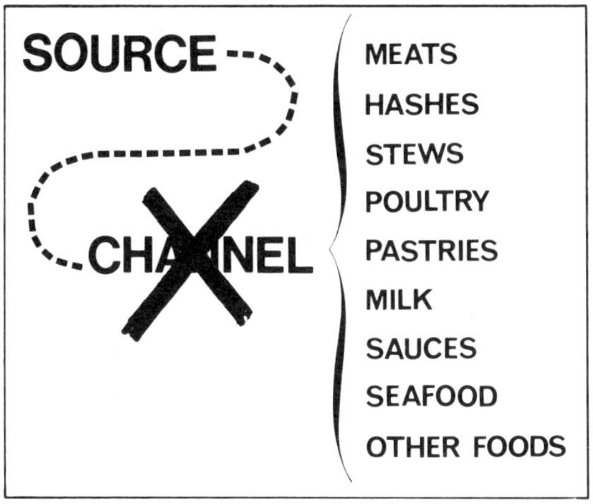

This is an illustration showing that bacteria, since they have no legs, arms or wings, must be transmitted by one of the channels (foods) listed at right. The foods listed are among those involved in food poisoning. These channels can only be X-ed out by the application of sanitary practices that prevent the spread of bacteria to food or people.

Foods Commonly Involved in Staphylococcus Food Poisoning—Staphylococci bacteria must have a suitable medium or food in order to grow and produce toxins. In many cases the food is high in protein and has been previously cooked. Often such foods are leftovers. Foods commonly involved include; cooked ham and other meat products; cream-filled pastries; potato, meat, poultry and shellfish salads; cheese; milk; chopped food; poultry dressings; bread pudding; sauces and gravies; hollandaise sauce; high protein, leftover foods.

A study of the above list reveals that many of these items are among those which workers in food preparation areas tend to handle more than is really necessary. Using the hands instead of the proper utensils available for mixing purposes is often done, with the excuse that it is faster, more thorough, but above all, easier. It is also easier to transfer staph (a commonly used abbreviation) organisms to food when improper preparation methods are used.

Most acid type foods are not suitable for the growth of staphylococci bacteria. On the other hand, salt and sugar solutions in strengths high enough to stop the action of many bacteria, have no effect on staph organisms. They are resistant to a 10 percent salt solution. This is the reason these germs have been found in cured meats.

Symptoms of Staphylococcus Food Poisoning—The presence of staphylococcus toxin within the body may be felt within one-half to 7 hours, with the average considered 2 to 6 hours.

Illness comes suddenly. Large amounts of saliva are noticed in the mouth. The sick individual feels nauseated, vomits, has abdominal cramps and diarrhea with resulting weakness. There is usually no fever. In some cases however, both blood pressure and body temperature may fall enough below normal to produce the condition known as shock.

The onset time will depend on how much toxin has been consumed and to what extent the person's system is resistant to it. The illness usually lasts from 1 to 3 days. The great majority of people recover completely, though some cases of death have been recorded.

Foods classified as potentially hazardous (such as those listed as being involved in staphylococci food poisonings) and which are not to be immediately consumed should be held at 45°F. or below, or 140°F. or above. The temperature range between 45° F. and 140°F. is often termed the danger or incubation zone, since bacterial growth and toxin production occurring at these temperatures can render food unsafe to eat.

Refrigeration Temperatures and Food—The mere action of placing a potentially hazardous food in the refrigerator immediately after cooking is not enough to ensure its safety, for the following reasons:

(1) Refrigeration temperatures can be raised (i) by overloading the facility; or

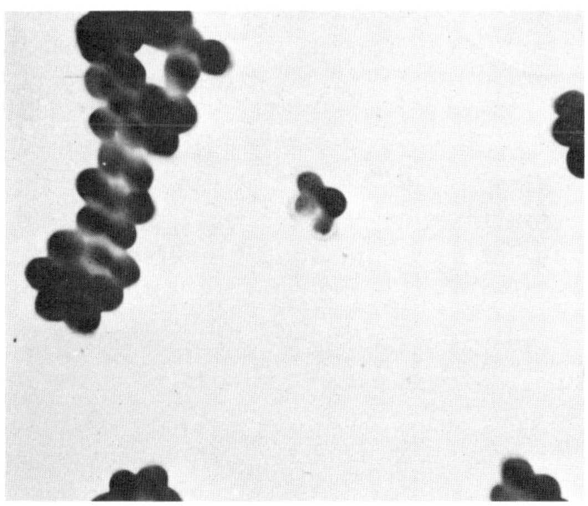

Staphylococcus aureus

(ii) because the cooling system requires adjusting; or (iii) by personnel constantly opening the refrigerator doors and thus admitting warm air.

An indicating thermometer accurate to within 2°F., easily readable but shielded to minimize breakage, should be installed in the warmest area of the refrigerator. Foodservice personnel who use the refrigerator frequently may note a thermometer reading closer to 60°F. than a desirable temperature of 40°F.

(2) It takes time for food to chill to the center of the food mass and several factors can influence the cooling rate. Foods stored in containers of stainless steel (which is a good conductor of heat) cool more rapidly than food stored in crockery, glass or plastic.

Food refrigerated in shallow pans cools faster than food stored in large, deep containers. If the type of food permits stirring (cream-filled pies, for example, could not be stirred without spoiling their appearance), a frequent stirring while cooling would hasten the chilling process by exposing more warm food surfaces to the refrigerator's cool air.

(3) Cream or custard filled pastries (puffs, eclairs, etc.) cannot be considered sufficiently cooled because they have been in the refrigerator an hour or so before serving. The pastry shells containing these food items act as insulation to hold the heat inside the shells.

For example, one noted investigator found that it took 3 hours or more to lower the temperature of the custard inside eclair shells from 80°F. to 50°F., although the refrigerator's temperature remained constantly between 38° and 42°F.

For this reason, health authorities recommend that such custard-filled pastries, after temporary refrigeration, be rebaked for 30 minutes at 375°F. before being served. Baking for this period of time and at this temperature does not affect the flavor or appearance of the food but does kill any staphylococci present in the products.

Rebaking of custard-filled eclairs at even higher temperatures (420° to 428°F.) for 15 minutes is also said to have no effect on taste or appearance of the bakery goods.

Experiments conducted by qualified investigators indicate that while the staphylococcus enterotoxin is extremely resistant to heat, it does become gradually less powerful after long exposure to very high temperatures such as in autoclaving. Such exposure, however, would render the foods unfit for human consumption.

It is obvious, then, that the control of staphylococcus food poisoning must be based on the control and destruction of the staphylococci organisms, before they can multiply in sufficient numbers to produce enough enterotoxin to cause illness to those who eat the foods concerned.

(4) Acting on the recommendation of experts, management in several areas now follows the practice of rapidly chilling cooked foods in shallow pans in the freezer rather than in the refrigerator.

As an example, it was found that a one-half gallon container of potato salad, with a temperature of 65°F. in the center of the salad mass, was lowered to a safe internal temperature of 41°F. in 35 minutes when placed in the freezer. The same amount of potato salad at 65°F. took 80 minutes to reach an internal temperature of 46°F. in the refrigerator.

Once the temperature has reached a desired level in the center of the mass, the food can be quickly transferred to a refrigerator if service time has not yet arrived. This, of course, assumes the refrigerator's temperature is maintained at the required degree of coldness.

(5) Sandwiches have been involved in staphylococci food poisonings because: (i) they are often prepared far ahead of serving time; (ii) their fillings may contain potentially hazardous foods such as ham salad, egg salad, etc.; (iii) refrigeration to a safe temperature is interfered with because the bread acts as an insulation.

The quick chilling of sandwiches has been recommended, provided they are stacked in the freezer in layers of two. Higher stacking prevents quick chilling, even in the freezer, because of the encasing bread.

Heating and Food—Thorough cooking of perishable foods requires temperatures high enough to kill most non-spore forming microorganisms, which include staphylococci. (A spore is a growing form of certain microorganisms which has an outer, protective covering that makes the germ cell able to withstand high temperatures.)

The Public Health Service Food Service Sanitation Manual gives 140°F. as a safe temperature for the holding of hot foods for relatively short periods of time. This holding temperature prevents the growth of staphylococci bacteria and the formation of toxin.

The problem with staphylococci bacteria, then, is to make sure the temperature of thoroughly cooked food does not fall below 140°F. for any great length of time before or during serving periods (as on cafeteria line steam tables or in a water bath or bain marie in a kitchen).

Again, it is the temperature in the *center* of the food mass which must be tested for the accuracy of the holding temperature rather than that of the outer surface. Whereever possible, the frequent stirring of the food as it is being cooked and while it is being held will expose all food surfaces to required heat, help ensure thorough cooking and safe holding temperatures.

Referring to the list of potentially hazardous foods, much research has been done on the proper cooking of stuffed turkeys. At an oven baking temperature of 300°F. investigators proved that food-poisoning organisms grow and multiply during the first stages of roasting, but that an internal temperature of 165°F. rids the stuffing of the living organisms.

If the cooked bird stands at room temperature for 20 minutes after baking, the temperature in the center of the stuffing continues to rise during that period. When frozen stuffed birds are roasted, it often takes twice as long for the internal temperature of the stuffing to reach 165°F.

Great care must be taken to observe correct time and temperature instructions for the roasting of large (over 18 lbs.) birds to ensure that the stuffing reaches the required safe internal temperature. Frequent testing with a thermometer is necessary.

Many food establishments bake poultry stuffing separately in shallow stainless steel pans. Poultry stuffing, meat loaf, croquettes and similar foods containing bread or cracker crumbs should be prepared only in amounts intended for immediate use unless the temperature in the center of the food mass reaches more than 150°F. during the cooking process. The bread and cracker crumbs create air pockets which act as insulation against heat.

Foodservice workers are often unwittingly responsible for holding cooked food at less than 140°F. This may happen when:

(1) Leftover foods are warmed rather than heated thoroughly to the center of the food.

(2) Food is placed in warm ovens at incubation temperatures to keep it heated.

(3) Food is not removed promptly from turned-off steamtables.

(4) Steamtable temperatures are turned down because personnel do not enjoy working in the heat.

Conclusions

(1) The only sure means of finding out whether food is sufficiently cooled or heated is to insert a thermometer into the center of the food mass.

(2) It takes *time* for the interior of the food to reach the proper cold or hot temperatures. Foods vary in the length of time it takes them to cool or heat. The length of time foods remain in the temperature danger zone between 45°F. and 140°F. has a direct relationship to the multiplication of any staphylococci present in the food and the production of their toxin which causes illness. Therefore, observance of correct time and temperature is an all-important factor in the control of staphylococcus food poisoning.

(3) Leftover cooked food requires thorough heating before reuse, and should not be kept long in any case. The length of time cooked protein foods are kept at temperatures between 40° to 45°F. and 140° F. will undoubtedly have a definite bearing on whether or not such foods will be safe to eat.

(4) The toxin produced by food-poisoning staphylococci is resistant to both heat and cold. Therefore, the bacteria themselves should be destroyed by heat *before* they have an opportunity to produce toxin. A temperature of 165°F. will kill the staph organism, but boiling (212°F.) of question-

able food will *not* destroy toxin already produced.

Refrigeration and quick chilling have already been discussed at length. However, if toxin has already been produced in foods, subsequent refrigeration has no effect on it.

Control of Staphylococcus

(1) Foodservice personnel with infected cuts, etc., colds or diarrhea should not work at food preparation. Good personal hygiene practices pertaining to handwashing and the other habits discussed under Communicable Disease should be followed routinely.

(2) Food

Since staphylococci organisms are literally everywhere, even correct personal hygiene habits of personnel cannot guarantee 100 percent protection for the foods in which these bacteria multiply and produce toxin. The presence of this poisonous material cannot be detected by the taste, odor or appearance of food, since staphylococcus toxin is tasteless, odorless and colorless.

Therefore, the wisest course to follow is to assume that the staphylococci organisms are already present in food and to take steps to keep them from multiplying and forming toxin. The sanitary rules and procedures relative to quick chilling, refrigeration, thorough heating and proper reheating of foods should be practiced by foodservice personnel under careful supervision.

(b) BOTULISM—This is a type of food intoxication involving improperly preserved food. The illness is caused by the toxin produced by the rod-shaped bacterium called Clostridium Botulinum. Six types or strains of this bacterium are known. Each type produces a different toxin. Four of the six types—A, B, E and F—are the principal causes of the disease in man.

Botulism germs are wide-spread in soil. Tests of soil samples from all over the world have shown that they can be found everywhere. Soil and dust are undoubtedly the source of contamination by these bacteria for vegetables. The organisms have also been found in the intestinal tract of animals, and this could be the source of the poison in meat or meat products which has caused botulism in human beings.

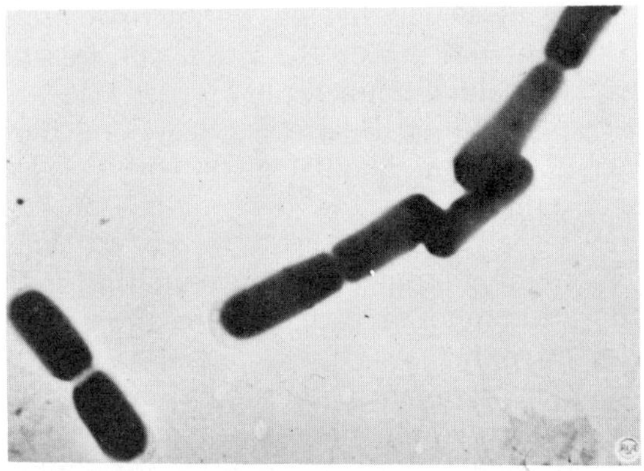

Clostridium botulinum

The toxin is extremely poisonous. It is so strong that two glassfuls, it is estimated, could kill the entire population of the world. At one time it was one of the seven agents revealed to be in the U. S. Army's arsenal of biological weapons.

The death rate from this illness is very high—23 percent of cases where the strains could be identified, much higher where the strains were unable to be identified prior to death.

Clostridium Botulinum is a spore-forming bacillus. The spores are extremely heat resistant. These spores are harmless in themselves, but under favorable conditions they can develop into active cells. It is these cells which produce the toxin.

Thus, in the preservation of foods, if the spores are present on the fresh food and are not killed during the preservation treatment, they can develop into active germ cells which multiply and produce toxin.

This is the reason botulism has occurred much more often in home canned foods than in those commercially canned. The scientifically developed and carefully controlled heat processes used by commercial canners have resulted in practically 100 percent safety of commercially canned products.

If, however, cans are damaged, leak, bulge, or are sprung, the contents should be regarded as unsafe.

Foods Involved in Botulism Poisonings—Home canned foods including fruits, vegetables, sausage and meats have been responsible for botulism outbreaks. Foods low in acid such as string beans and beets have been involved.

From 1950 through 1969 there were 182 outbreaks of botulism reported to the United States Communicable Disease Center. These outbreaks involved 415 cases with 127 resulting deaths.

Until a few years ago, outbreaks of botulism were most frequently caused by type A or B toxins and were usually associated with ingestion of home-canned vegetables, fruits or meat products. Botulism due to type E toxin, although recognized as having occurred in the United States as early as 1932, was not recognized as a major problem until 1963, when 22 cases of this type were reported: 16 of the 17 outbreaks of type E botulism have been traced to fish or fish products.

Symptoms of Botulism—The Clostridium Botulinum toxin affects the central nervous system. Because of this it is called a neurotoxin. Double vision occurs early, as do headaches and a general feeling of tiredness, muscular weakness and difficulty in swallowing. Breathing becomes irregular and if death occurs, it is usually due to respiratory failure. The symptoms usually appear 12 to 36 hours after the toxin has been consumed, although in some cases 6 to 14 days have elapsed before illness begins.

Control of Botulism Poisoning

(1) Boiling for 15 minutes, while constantly stirring the food, will destroy the botulism toxin.

(2) Pressure cooking in canning operations.

(3) Proper curing of meat.

(4) The addition to the food of certain suitable and approved acids which will prevent or retard the growth of the bacteria.

(c) CLOSTRIDIUM PERFRINGENS (WELCHII)—This organism is a spore forming bacillus that thrives in the absence of air. As in the case of botulism, there are several types of this bacteria. Only two of them, types A and C, cause food poisoning. Type A also causes gas gangrene.

Clostridium perfringens has rather high nutritional needs, requiring over a dozen amino acids (the units that make up proteins), and many other growth factors as well. These nutritional needs probably play a role in the kinds of foods in which it is found. Foods associated with outbreaks caused by this organism are usually high in protein—for example, meats and gravies, poultry and poultry stuffings which have not been thoroughly cooked and which have been allowed to cool slowly. The best temperature for growth of C. perfringens is between 109° and 117°F.

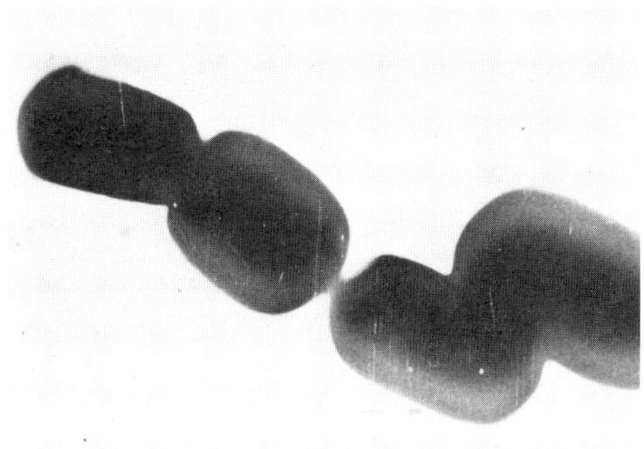

Cells of Clostridium perfringens in various stages of division

In the United States, during 1966 to 1968, C. perfringens foodborne illness accounted for 67 reported outbreaks. During 1966 and 1968, C. perfringens was responsible for more reported cases of foodborne diseases than any other agent.

Foods involved in the outbreaks of Clostridium perfringens foodborne illness are usually meat or poultry that has been boiled, stewed, or lightly roasted; or meat and poultry stews, sauces, gravies, pies, salads, casseroles and dressings. The food involved is invariably held at room temperature or refrigerated in large masses for several hours, often overnight or longer. Outbreaks sometimes follow banquets or meals prepared at hospitals and schools where large amounts of meat and poultry are involved. C. perfringens is a disease closely associated with the foodservice industry.

When sufficient numbers of the bacteria are taken into the body, diarrhea and abdominal pain, accompanied by large volumes of gas in the intestine, occur after an incubation period of 4 to 22 hours.

C. perfringens is a normal inhabitant of the intestinal tract of man and animals. It is found in the soil in both spore and active forms. It is frequently found in dust from kitchens.

The only practical method of controlling this bacteria, since it is so widespread in nature, is to attempt to prevent it from growing to large numbers in prepared food. This involves the thorough cooking of foods (especially meats); the practice of sanitary measures in the establishment; strict attention given by foodservice workers to good personal hygiene habits, in order to prevent recontamination of the food; the rapid chilling of cooked foods not eaten while hot or held at 140°F. or above, and the refrigeration of such foods at 45°F. or below (refrigeration methods are discussed in Food Protection); special attention given to leftover meat and poultry items. These cooked foods should be refrigerated at constant temperatures of 45°F. or below and thoroughly reheated to internal temperatures of 165° to 212°F. immediately before serving.

2. Chemical Food Poisonings

Harmful chemicals are sometimes introduced into foods accidentally and illness is the result. These chemicals could well be required in the sanitary maintenance of the operation and might include cleaning compounds, bactericides, insecticides, rodenticides, etc.

Symptoms of the poisoning may appear within a few minutes or over a longer period of time, depending on how much of the poison has been taken into the system and how toxic it is. The symptoms may include vomiting, diarrhea, cramps. With some chemicals, dizziness and signs of paralysis appear. Convulsions may occur.

Control of Chemical Food Poisons

The United States Public Health Service Food Service Sanitation Manual lists the following suggestions for the control of chemical poisonings:

(1) Insecticides and rodenticides in powdered form should be colored some distinctive hue to prevent their being mistaken for foods such as flour, powdered eggs, etc.

(2) All containers of insecticides, rodenticides, etc. should be distinctly marked or labeled so their contents can be easily identified.

(3) Storage should be in cabinets used for no other purpose, or in a place away from the food preparation, food storage or clean equipment and utensils storage areas.

(4) Bactericides and cleaning compounds should not be stored in the same cabinet or area of the room with insecticides, rodenticides or other toxic materials.

(5) Bactericides, cleaning compounds or other compounds intended for use on food-contact surfaces should not be applied so that any remainder of toxic material is left on such surfaces or any hazard caused to employees or customers.

(6) Pesticides or any other poisonous materials should not be used in a way that might create contamination of food, equipment or utensils, or cause any other hazard to foodservice workers or customers.

3. Metallic Food Poisonings

Acid liquid food or drink prepared in cadmium plated food containers; foods cooked in inexpensive enamelled cooking utensils containing antimony; certain foods prepared in galvanized iron kettles containing zinc—all these have resulted in vomiting, nausea and abdominal pain a few minutes to an hour after such food has been eaten.

Toxic metallic containers should not be used for the preparation or storage of acid foods. The availability of modern, non-toxic cooking utensils, designed for today's needs, has practically eliminated the danger of this type of food poisoning occurring in commercial establishments. It is more likely to happen in the private home where a favorite older cooking utensil is still in use, or where the facts concerning acid foods prepared or stored in toxic metal containers may not be known or understood.

4. Food Poisoning Due to Natural Causes

Fish Poisoning—Six general types of toxins have been found in different species of fish inhabiting tropical and sub-tropical waters. The species involved include snapper, bonito, moray eel, blow fish, puffer, barracuda, etc.

Many of these fish are extremely poisonous only at certain times of the year and it is believed the toxins come from the diets of the fish.

The symptoms of this illness may occur in less than an hour or not for several hours after the fish has been eaten. Symptoms include nausea, vomiting and abdominal pain, blueness of lips, tongue and gums, loss of taste and—in fatal cases—respiratory paralysis. There may be a death rate of 60 percent in severe attacks.

The only control method is to avoid eating any of the poisonous species.

Shellfish Poisoning—This type of poisoning occurs when shellfish, such as clams, feed on certain types of marine organisms (plankton) producing "red tide." The toxin in the plankton involved is one of the strongest known and is in the same class of poisons as strychnine. Occurrences of "red tide" where the water is stained red in daylight and glows with light in the darkness are familiar to residents of many coastal areas of the United States.

Some of the symptoms of shellfish poisoning are numbness of the lips, loss of muscular power in the neck and legs, drowsiness and respiratory paralysis. Death from respiratory failure has occurred in about 9 percent of the illnesses, but if the patient lives for 12 hours after onset of symptoms, his chances of recovery are good.

Onset of symptoms may occur from 5 to 30 minutes after food is consumed.

Commercial harvesting of shellfish is carefully controlled and approved by the State shellfish authority. If the source is outside the State, it must be certified by the State of origin.

Clams or mussels of unknown quality should not be eaten.

Poisonous Plants—Plant food poisoning is not apt to be a problem in the commercial food establishment, hence will be mentioned only briefly for informative purposes. The research and comments of health authorities should be of interest—and help—to food-service workers who are tempted to eat berries, greens, mushrooms, etc. without knowing for sure what these foods may contain or whether they are actually eating what they believe they are.

Everyone has heard stories of the amateur mushroom hunters who have, by mistake, picked poisonous toadstools and eaten them. Illness, sometimes death, results.

Water hemlock grows in wet places and is quite common throughout the United States. Its roots have been mistaken for wild parsnips and the plant for horseradish.

Rhubarb leaves may contain 0.4 percent oxalic acid. Berries of the deadly nightshade have caused death. Children who have mistaken mountain laurel for wintergreen have been made seriously ill. There have been many other cases where mistakes in identifying a food item correctly in its natural state has resulted in disaster.

The average citizen who enjoys mushrooms should eat only those produced commercially. As for berries and other types of food which might be suspect, only those *known* to be safe should be eaten.

5. Physical Cause of Food Poisoning

Since we live in a nuclear age, the possibility of future contamination of foodstuffs by materials used in the manufacture of atomic weapons (example, strontium 89, 90) should be considered.

Such contamination could be caused by fallout, radioactive wastes, etc. and could affect many foods in the vegetable, fish, meat, cereal and milk categories. The results of eating this contaminated food would probably not be evident for many years.

Sufficient amounts of fallout on the ground could contaminate plant roots and foliage as well as the blossoms of fruit trees. Ripe fruit would still contain the poison.

Animals grazing on contaminated grass would take these toxic materials into their bodies.

Thus, beef and dairy cattle and other animals used as food could be affected by these materials.

Fallout materials in the sea could be absorbed by plankton which, in turn, would be eaten by shellfish and small fish. The larger game fish feeding on the smaller varieties, could carry the contamination to human beings who ate them.

Aside from stopping tests of nuclear weapons, a present control method of constant checking to determine the levels of radioactive fallout, wastes, etc. should be—and will be, of course—continued. At the present time, the amount of toxic material released into the atmosphere by atomic activities is well below the danger level.

Research workers in this field are investigating the many areas of causes and effects involved in this problem so that, if ever necessary, the proper steps can be taken to cope with it.

FOOD INFECTIONS

These are the illnesses due to contamination of food by any one of numerous types of organisms which gain entrance to food, grow and multiply–under favorable conditions–to dangerous numbers and are eaten with the infected foods.

1. Bacterial.

While many examples could be given of bacterial food-borne illnesses, only a few of the most important will be discussed here.

The general symptoms of a number of bacterial food infections are the same. They include headaches, chill, abdominal pain, diarrhea, nausea, vomiting, muscular weakness, faintness and thirst.

These illnesses are usually transmitted through foods infected and allowed to remain for some length of time at room temperatures suitable for the growth of the infective organisms. This is particularly true of those foods which require considerable handling in their preparation, such as hash, salads, chopped meats, cream and custard-filled pastries and pies.

Foods most commonly involved in bacterial food infections include: custards, milk, cream, ice cream, sea foods, meats, eggs, meat products, poultry and poultry dressings, shellfish, raw salads and raw vegetables, salad dressings, bread pudding, cream pies, eclairs and any filled pastry.

(a) SALMONELLOSIS–There are many types of salmonella bacteria. The organisms are found all over the world, infecting both man and animals. They are common in domestic poultry, livestock and rodents. The intestinal tracts of animals and human beings provide naturally suitable conditions for the existence of salmonella.

Knowing this, the foodservice worker can understand why salmonella infections are involved in many cases of illness originating in foodservice operations. The bacteria may be present in under-cooked, semi-cooked or raw foods which have been exposed to animal wastes.

Workers who do not wash their hands thoroughly after restroom visits or after

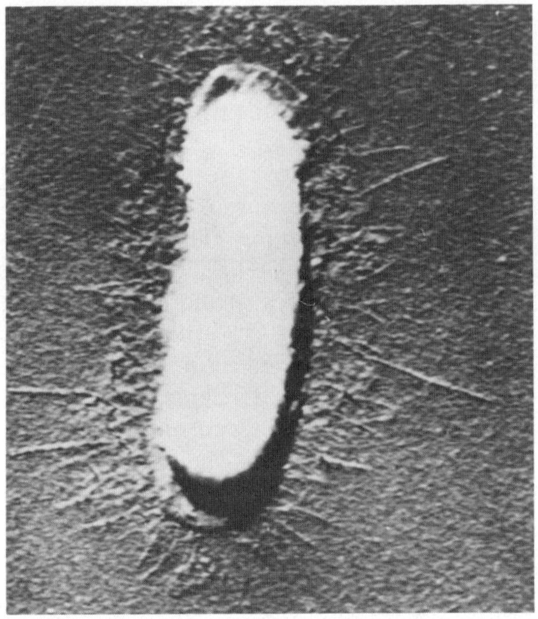

Salmonella typhimurium

handling raw poultry (especially if it is cleaned on the premises), raw meats and raw eggs may contaminate any other food or utensils contacted.

The foodservice worker himself may be a carrier, since the germ may be present in his intestinal tract–not, perhaps, in sufficient numbers to make him ill, but still capable of being transmitted to food through careless personal hygiene habits.

Mice, rats or cockroaches may contaminate food by dragging filth over food and food utensils, or by intestinal deposits brushed off into food or containers.

The foods most often involved in salmonella infections are meat, poultry and eggs, or salads made of those ingredients.

The symptoms may appear 12 to 24 hours after eating the infected food. The average time is 18 hours.

Control of Samlonella Infections

(1) Thorough cooking throughout the food mass will destroy salmonella.

(2) Proper refrigeration will prevent the multiplication of these organisms.

(3) Do not keep foods at incubation temperatures (between 45° and 140°F.).

(4) Observe correct hand habits, especially after restroom visits and contact with possibly contaminated foods.

(5) Protect foods from rats, mice and insects.

(6) Follow required sanitary food preparation methods.

(7) Follow correct dishwashing routines and proper sanitizing treatment.

(8) Use pasteurized milk and milk products as well as meats and poultry processed according to sanitation regulations.

(b) BACILLARY DYSENTERY

The germs which cause this illness are found in the intestinal discharges of infected persons. It is transmitted through contaminated foods or water, by flies and, occasionally, by carriers.

Moist or mixed raw foods are the items commonly involved.

Control of Bacillary Dysentery—Control measures include the standard sanitary rules of thorough cooking, proper refrigeration, correct sanitary methods of food preparation, use of pasteurized milk and milk products, good housekeeping, protection of food from flies, water protection and treatment and sanitary sewage disposal.

(c) TYPHOID FEVER.

This is an illness transmitted through food, milk, shellfish or water supplies which have become polluted with the body discharges of a person infected with the germs of this disease. It may also be spread by flies which carry the typhoid bacilli from soiled articles or sewage to foods, fishes and utensils. Typhoid carriers play an important role in the transmission of the disease.

The foods commonly involved in typhoid fever are milk, raw salads, shellfish, or any high protein food which has been contaminated with the disease-causing organisms of this illness.

Onset of the disease may be from 7 to 28 days.

Control of Typhoid Fever—Control measures are the same as those listed for bacillary dysentery. In addition, typhoid carriers should not work in foodservice; shellfish should be purchased from approved sources. Recently, protection by immunization with vaccine, now widely used, has resulted in a marked decrease in the number of reported cases of typhoid fever.

Para typhoid Fever is a food-borne illness very similar to typhoid and is caused by certain bacteria of the Salmonella group. General symptoms and control measures are the same as for typhoid fever. The disease itself is much milder than typhoid. In food outbreaks of para typhoid, the symptoms usually appear in from 1 to 15 days.

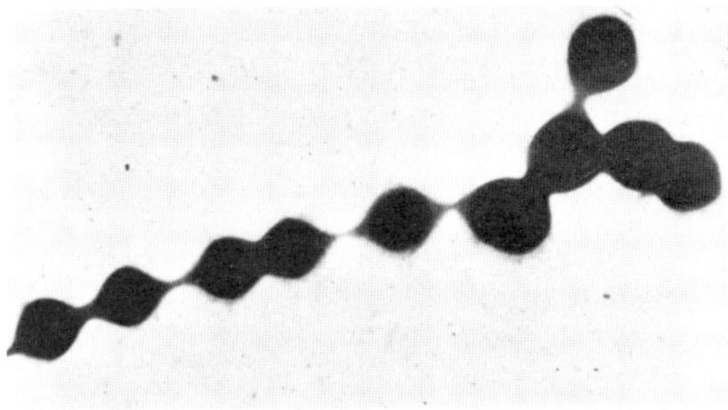

Streptococci

(d) STREPTOCOCCAL FOOD INFECTIONS

The streptococcus organism may be found in the intestinal tracts of man and animals, in discharges from the nose and throat, and in wounds. Hence, food may be contaminated by soiled hands, by droplet infection due to coughing, sneezing, or even talking. Since germ-laden droplets can be expelled for quite some distance through the air (refer to Communicable Disease discussion), floor dust raised by sweeping and breathed in by personnel or others must also be considered a transmission possibility.

Turkey, chicken, eggs (and salads containing these food items), milk and gravies are frequently involved in streptococcal infections.

The first symptoms may appear from 2 to 18 hours after the contaminated food has been eaten.

Control measures include prompt and adequate refrigeration, thorough cooking of food, careful handwashing and other necessary personal hygiene habits, pasteurization of milk and milk products and strict attention to the oft repeated rules of sanitary food preparation.

Septic sore throat and scarlet fever are also streptococcal infections.

(e) COLIFORM FOOD INFECTIONS

Members of the coliform group of organisms are present in the intestinal tract of mankind. They may give rise to severe diarrhea, abdominal pain, nausea and vomiting, especially in infants. The onset of the illness is rapid, from one-half to 3 hours.

Milk is the food involved, though a polluted water supply may also be responsible.

Control of Coliform Food Infections—In a foodservice operation, thorough handwashing after restroom visits is a must for personnel. Tests for personnel carriers should also be made. The required washing and sanitizing routines for dishes and utensils are very important. Health officials will check for polluted water supplies or defective sewage systems, and see that corrective measures are taken when necessary. The fly population in the establishment should be controlled.

2. Food Infections Due to Parasites

A parasite is a plant or animal which obtains its nourishment from another living organism.

(a) AMOEBIC DYSENTERY

One-celled animals known as amoeba are responsible for this disease. They eat the red blood corpuscles of the body and also the cells which line the intestine. Amoebic dysentery is transmitted through contaminated foods, especially those that are commonly served cold and moist; flies; raw vegetables such as celery and lettuce which have been fertilized with excreta; and chronic carriers. The onset of symptoms may be from 5 days to a few months, but is more often from 3 to 4 weeks.

Control of Amoebic Dysentery—The personal habits of foodservice personnel are an important factor (again, thorough handwashing must never be neglected); thorough cooking of food; water protection and treatment; sanitary sewage disposal; and careful, thorough washing of raw vegetables.

(b) TRICHINOSIS

This disease is caused by eating infected meat—usually pork—which has not been cooked enough to kill the larvae (worms) which cause the disease.

The immature worms are ingested in undercooked pork, if such is infected. The worms pass through the stomach to the intestine, where they grow into adult worms and mate. The young worms are microscopic and are born in the intestine. They burrow into the blood vessels and are carried to the various muscles in the body, into which they burrow and wriggle around for some time.

It is this working around in the muscles which causes the pain and cramps associated with Trichinosis. If the eye muscles are involved, blindness can result. Occasionally, the brain is attacked by these larval worms and death results.

The hogs may obtain the worms by eating garbage containing pork scraps that have not been cooked and which contain trichina larva.

Trichinosis seems to be most prevalent in the very heavily populated eastern states, where it is just as high in cities as in the rural regions. Autopsies performed in this area reveal that about 27 percent of all the people have had the infection during their lifetimes. The average for the country as a whole is 17 to 18 percent.

Control of Trichinosis—Control measures include the thorough cooking of pork; cooking garbage at 212°F. for 30 minutes; freezing; and the elimination of rodents from hog lots.

(c) INFECTIOUS HEPATITIS

Although the fact that the infectious hepatitis virus is found in human feces (intestinal waste) suggests that this virus might be transmitted in foods, actual reports of food-borne outbreaks began to appear with regularity only in recent years. This is probably because, in the past, the incidence of infectious hepatitis has been small. However, with an increase in population and a consequent increase in sewage pollution, the incidence of this disease has risen.

The first food-borne outbreak of infectious hepatitis reported in the United States occurred in Mississippi and Alabama during the early months of 1961. Most of the cases involved the eating of raw oysters which had been harvested from an oyster bed located about a mile below a community raw sewage discharge.

At about the same time, 459 cases of infectious hepatitis were attributed to raw clams harvested from the Raritan Bay in New Jersey. From that time until the present, many other outbreaks have been traced to raw oysters or clams that were harvested from contaminated waters.

Other food-borne outbreaks of infectious hepatitis have been traced to orange juice, roast pork, green salad; luncheon meat, frozen strawberries, glazed doughnuts, iced pastry, sandwiches, whipped cream cakes, and mayonnaise in potato salad. These foods were contaminated by infected workers who practiced poor personal hygiene habits and then touched foods.

An outbreak in a high school cafeteria in Missouri City, Texas in 1959 was food-borne in only a limited sense. A worker, who did not handle food, apparently contaminated cafeteria trays on which food was served. Twenty cases of infectious hepatitis resulted from this contamination. This incident is mentioned to show the possibility of spreading this disease not only by the foodservice worker directly involved with preparation and serving, but also by personnel who may not actually handle the food.

The majority of foods implicated in this disease have been cooked very little or not at all, as in the case of raw shellfish previously mentioned.

One notable exception was the roast pork incriminated in the Bushy Hall, England, outbreak in 1963. The pork had been roasted thoroughly, cooled, and boned by hand by a cook who showed signs of jaundice (a yellowing of the skin which usually accompanies infectious hepatitis). This same cook subsequently sliced and warmed the meat on a steamtable before serving.

Thus, thoroughly cooked food may become a hazard if improperly handled just before serving, or if it is served on a contaminated dish or tray.

The thorough cooking of foods is an important measure in the control of infectious hepatitis. Necessary also is the acceptance by foodservice workers of the highest possible standards of personal hygiene. Any individual showing signs of jaundice should consult a doctor and should not return to work without clearance from a physician.

Responsibility of Personnel for Introduction of Harmful Bacteria

Personnel may be responsible for the introduction of harmful bacteria which cause food intoxication and food infection in any of the following ways:

(a) By failing to wash vegetables properly. All particles should be separated and scrubbed until no filth-containing bacteria may remain.

(b) By being carriers of infection.

(c) By contaminating food with unclean hands due to failure to thoroughly wash the hands after restroom visits

(d) By contaminating food by working with boils, sores, etc. containing pus which teems with bacteria and which will multiply like wildfire in foods.

(e) Contamination of foods by waiters or waitresses who handle butter, bread, doughnuts, etc. with the hands instead of using tongs or a fork.

(f) Personnel who make sandwiches should not place dirty plates or other utensils on the sandwich preparation table. Towels used to wipe containers should not be used to wipe this table.

(g) The refrigerator should be thoroughly cleaned and sanitized weekly. No spoiled foods should be left in the refrigerator.

(h) All pork should be thoroughly cooked. Raw and rare pork should always be avoided. Masses of food like spaghetti should be stirred well so that cooking will be complete in the center of the mass.

(i) Milk should be dispensed always in accordance with health department regulations.

(j) Commercially prepared sandwiches should be kept under refrigeration from the time of preparation until the time of service.

(k) Ice for drinks should be handled with scoops or tongs. Most food-borne disease outbreaks have been due to three things:

(1.) Food prepared far in advance.

(2.) Poor refrigeration.

(3.) Careless handling.

Construction and proper operation of appliances, etc. will be discussed under Food Protection.

SUMMARY

The fact that management in foodservice operations are not aware of any food-borne illnesses having originated in their establishments does not mean that this has never happened. The symptoms may begin several days or longer after infected food has been eaten. And unless large numbers of people become ill, usually very suddenly, the persons concerned are likely to blame their sickness on some other cause, such as food eaten at the latest meal in their own homes.

When large groups are involved, the press, radio, and television are quick to use the incident as news and, customer-wise, the unfavorable publicity to the establishment may be disastrous. In addition, some of the people involved may bring damage suits against management.

On a national scale, any reported occurrences of contamination found in commercially canned, processed or frozen foods never fail to bring unfavorable publicity to the companies concerned–even when such manufacturers or processors operate under strict sanitation regulations.

Over the past two or three years, the Food and Drug Administration has caused three major recalls and nation-wide alerts involving canned mushrooms. In one of these cases, FDA field investigators found evidence of botulism toxin and decomposition in the products. These conditions were blamed on a canning machine that packed the mushrooms too tightly into cans, not allowing enough space for sufficient cooking to destroy the bacteria and spores.

Because many of these large containers of mushrooms were distributed to processors of other food products using mushrooms in their recipes, more than 100,000 chicken dinners, frozen pizzas and tuna noodle casseroles were removed from retail grocery shelves. Over 13 million servings of a popular soup mix were removed from the market because they were contaminated with salmonella bacteria. The organism was traced to the manufacturer of the noodle soup used in the mixes.

In other instances, 334,000 pounds of popcorn, 24,000 nationally distributed frozen strawberry pies, and some 2000 cases of a name brand candy were recalled. All were contaminated with rodent or insect material or both.

These examples are given to show that employees in any area of a foodservice operation have a great responsibility to the public as well as to their employers. Not for one moment must awareness be relaxed or eyes shut to any apparent defect in machinery or operating methods.

During times of great pressure, the temptation to take short cuts, to close the eyes to sanitation errors is almost irresistible and, at the same time, understandable. Nevertheless, in the food industry such errors must not occur. The disregarding of just one step in a routine sanitation procedure or method may have far-reaching consequences for both the employee and the people who are made ill as a result of that error.

The total number of food-borne disease outbreaks per year in the United States is impossible to estimate. Health departments do not know how many actually occur in their areas because many people do not consult a doctor or report a suspected illness to health officials. Hence, health department records often list fewer cases than they would if every case of food-borne illness was identified for what it is.

Foodservice personnel, therefore, must keep pace in their methods with the efforts of equipment manufacturers, construction engineers, product manufacturers, food processors, health departments and all the others who have worked to bring the physical facilities, the tools, and the food itself to their present sanitation levels. The efforts of all those cooperating to raise sanitation standards are of little use, if the operational procedures of personnel continue to permit occurrences of food-borne diseases in their establishments because of personnel carelessness about following routines and their neglect of required personal hygiene habits.

RANDOM SELECTION OF FOOD-BORNE DISEASE OUTBREAKS

Persons Sick	Type Illness	Type Establishment	Food Vehicle
1700	Food Poisoning	Hotel Banquet	Dressing
430	Chemical Poisoning	Prison (Sabotage)	Gravy & Soup
145	Food Infection	Hospital	Ground Veal & Gravy
206	Food Infection	Veterans Banquet	Turkey
130	Food Poisoning	School	Mince Pie
124	Food Poisoning	School	Potato Salad
246	Food Poisoning	Private Home	Boston-Cream Pie
137	Food Poisoning	Hospital	Tongue
210	Food Poisoning	Banquet	Giblets, Gravy, Beans
212	Food Poisoning	School	Egg Salad Sandwiches
200	Food Poisoning	Wedding Banquet	Turkey Dressing
230	Food Poisoning	Institution	Butterscotch Pudding
1300	Food Poisoning	College	Veal & Dressing
55	Food Poisoning	Restaurant	Turkey A La King (Chef recently recovered from diarrhea)
70	Food Poisoning	Restaurant	Cream Pie
100	Food Poisoning	Restaurant	Crab Salad
122	Food Poisoning	Restaurant	Chopped Liver
245	Food Poisoning	Restaurant	Rice Pudding
101	Food Poisoning	School	Egg Salad–Bologna
650	Food Poisoning	College	Ham
400	Gastro-enteritis	High School	5 Food Handlers with positive nose and throat cultures. Several foods involved.
200	Gastro-enteritis	Bakery	Meringue Pie
150	Chemical Poisoning	School	Fluoride on Dish Shelves
388	Gastro-enteritis	School	Fricassee Chicken
200	Food Poisoning	Caterer	Gravy & Dressing

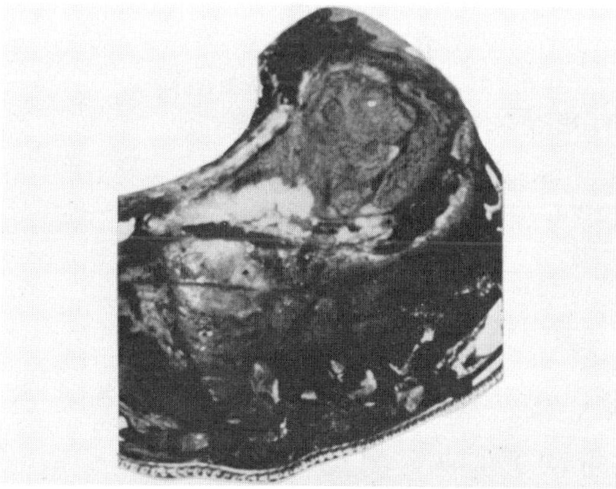

Sanitized utensils, careful handling, adequate refrigeration protect natural food flavor.

FOOD-BORNE DISEASE QUIZ

1. True or False (check True or False):

	True	False
(a) Food Poisoning and Food Infection mean the same thing.	( )	( )
(b) A toxin is a poisonous substance produced by certain bacteria.	( )	( )
(c) Illness due to toxins in food is called Food Intoxication.	( )	( )

2. Complete the following statements:
   (a) The most common of all food poisonings are ______________ and ______________.
   (b) Another type of food poisoning is ______________________.

3. Check correct statements.
   (a) If chemicals necessary for use in foodservice get into food, they can cause illness when the food is eaten. ( )
   (b) Pesticides should be kept where they are handy. ( )
   (c) The safe way is to eat commercially produced mushrooms. ( )
   (d) As long as any metal container is in good repair, food may safely be prepared or stored in it. ( )
   (e) All home-canned foods are wholesome and safe to eat. ( )

4. Fill in the following statement:
   Food Infection is caused by ______________________ which gain entrance to food.

5. Fill in the following statement:
   Three foods commonly involved in food infections are ______________, ______________, ______________.

6. True or False (check True or False):

	True	False
(a) Salmonellosis does not occur very often.	( )	( )
(b) Trichinosis is caused when improperly cooked, infected pork is eaten.	( )	( )
(c) Foodservice workers *must* wash hands thoroughly after every restroom visit.	( )	( )

7. Check correct statements.
   (a) Carriers may work in foodservice operations. ( )
   (b) Food should not be prepared long in advance of serving time. ( )
   (c) There is no need to refrigerate sandwiches. ( )
   (d) Food must be heated to the center of the mass. ( )
   (e) Left-over food, once cooked, need only be warmed up for a second serving. ( )

8. Fill in the following statement:
   Foods should not be held between ^{o}F. and ^{o}F. This temperature area is called the ______________ ______________.

9. Fill in the following statement:
   It takes _______ for food to chill to the _______ of the food mass. Rapid chilling in _____________ pans is possible if food is first placed in the ______________.

10. True or False (check True or False):
   (a) Health departments always know how many cases of food-borne illness occur in their jurisdictions. True ( ) False ( )
   (b) The carelessness of foodservice workers may lead to an outbreak of food-borne illness. True ( ) False ( )

5 Food Protection

During the discussion of the food-borne diseases, methods were mentioned that control the spread of those illnesses transmissible through foods when such food items are handled in an insanitary, improper manner. These methods are a part of food protection.

In addition to protecting food from disease-causing organisms, there is also the problem of food spoilage. This will be discussed here. Then the general rules for food protection will be listed for the students' guidance and easy reference.

The presence of quantities of disease-causing organisms or their toxins in food is realized only when people who eat the food become ill. Spoiled food, however, is usually easy to detect because of its appearance and odor.

Spoilage is the decay or rotting of foodstuffs due to the action of certain bacteria. These bacteria may spoil food by eating some of it and leaving their waste products in it; or they may produce toxins.

This type of bacterial action in food does not mean that eating such food inevitably results in illness. There are people who actually like certain foods which others may consider spoiled.

Take, for example, limburger cheese. While this food is relished by many individuals, the mere odor of limburger is enough to nauseate others. Rotten eggs are enjoyed by the Chinese, as are rotten seal meat and fish by Greenland Eskimos. (See *Food Poisoning,* by G. M. Dack, Ph.D., M.D.)

These examples, quoted because they are considered interesting, are not intended as guidelines for foodservice workers. The rule in foodservice operations is that any food considered unfit for use should be immediately discarded.

Food contaminated by disease-causing organisms or their toxins is "sick" food, just as people are often termed sick. Foodservice workers must learn how to prevent sickness in food, just as physicians learn how to control and prevent illness in human beings.

Food is also delicate in nature, susceptible to spoilage.

Therefore, two of the most important parts of the foodservice worker's sanitation training are: (1) how to prevent the spread of diseases transmissible through foods, and (2) how to prevent foods from spoiling.

A summary of the conditions required for disease transmission, as covered in Food-Borne Diseases, will provide a reference background for the sanitary controls and preventive measures to be suggested in this chapter.

1. The disease-causing organisms must be present in the food preparation area. Perhaps they are on the body or clothing of a foodservice worker; they might be found in floor dust or in the air, in or on the bodies of rodents or insects, possibly in meats (such as trichinae in pork), other meat products or poultry.

2. Whatever the source, the organisms may contaminate food by direct contact (transference) from the source to the food. Or there may be an indirect contact from the source to an object. (Review the Communicable Disease chapter for details on direct and indirect contact).

3. The food must be suitable for the growth and multiplication of the organism which has gained entrance to it.

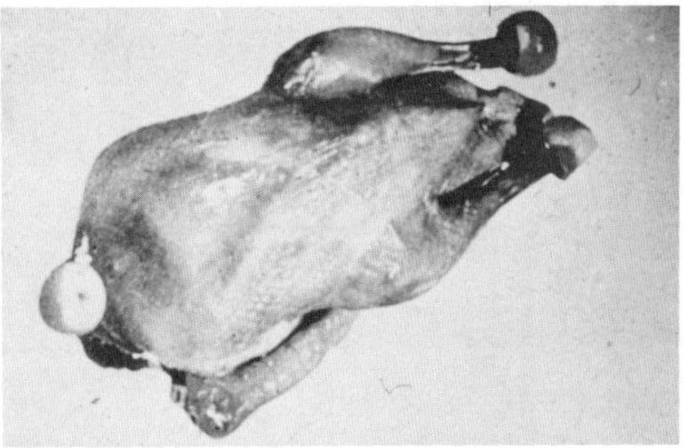

Temperature penetration–hot or cold–to center of large food mass requires time. DO NOT ALLOW TO STAND AT ROOM TEMPERATURE.

4. The temperature of the contaminated food must remain in the growth range (45°F.-140°F.) for a long enough time to allow multiplication of the organisms or their toxins to such quantities that illness will be caused in those who eat the food.

5. The food containing the disease-causing organisms on their toxins must be consumed in an amount sufficient to cause illness to the person eating the food.

The degree of illness could vary with the type and strength of the organism or toxin. The age of the person and his body's ability to resist the organisms' action also have an effect on how ill he will become.

When conditions necessary to the outbreak of food-borne disease can be prevented or controlled, such illnesses can be avoided in a foodservice operation.

The following list of suggestions provided by health authorities discusses how both harmful and food spoilage bacteria may be eliminated or controlled.

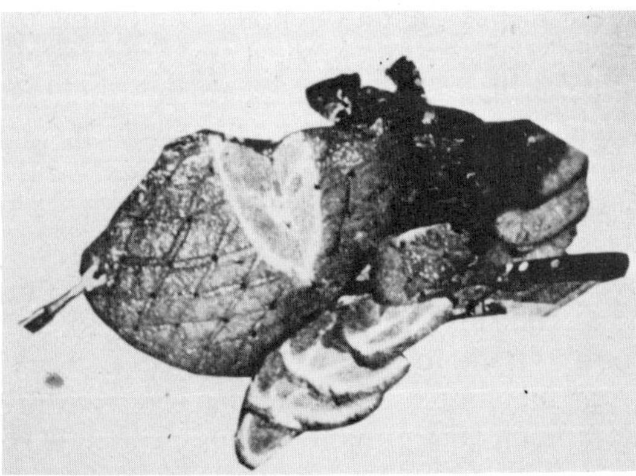

Thorough cooking, sanitized utensils, careful handling, adequate refrigeration–all are essential in food protection.

Foods in General

1. The U.S. Public Health Service Food Service Sanitation Ordinance and Code, page 40 states as follows:

"Food used in foodservice operations shall be from a source approved, or considered satisfactory by the health authority and which is in compliance with applicable State and local laws and regulations. Food from such sources shall have been protected from contamination and spoilage during subsequent handling, packaging, and storage, and while in transit."

2. Shellfish, meat, meat products, poultry and poultry meat products should be purchased from properly inspected sources. Such inspected foods are stamped on their containers or may have attached tags indicating the source's official approval.

3. Only pasteurized fluid milk and fluid milk products should be used or served. Dry milk products may be reconstituted (brought back to the original form) in the establishment if used for cooking purposes only.

All milk and fluid milk products used for drinking purposes should be served in the original individual containers in which they were packaged at the milk plant, or from an approved bulk milk dispenser.

Cream, whipped cream or half-and-half may be served from original one-half gallon containers or from an approved dispenser. Milk for mixed drinks requiring less than ½ pint may be poured from one quart or ½ gallon containers packaged at the milk plant.

Preventing Contamination of Food in the Establishment

1. Foodservice Personnel

Foodservice workers must learn and apply correct food preparation procedures as well as sanitary personal hygiene habits. These include:

(a) Wash and scrub all vegetables care-

AVOID THE INCUBATION ZONE

To be safe always hold foods below 45° or above 140°F.

fully. Pay particular attention to those intended for use in salads and those to be eaten raw. Residuals from chemical sprays may still be present on the vegetables.

(b) Storeroom personnel should examine canned food for bulges, dents, leakage, etc. before issuing to kitchen personnel. Opened cans with off odors or unusual appearance should be shown immediately to a supervisor.

(c) Custard or cream filled pastries should be immediately refrigerated, quick chilled first in freezer if there is not enough time for thorough cooling to center of pastry before serving. Test, wherever practicable, with a thermometer. If only temporary refrigeration is possible, health authorities recommend that custard filled pastries be rebaked for 30 minutes at 375°F. before being served.

(d) Refrigerate any sandwiches containing potentially hazardous food such as egg, ham, poultry salads, etc.

(e) Cook pork until well done (150°F. or above in center of mass. Test with meat thermometer.)

(f) Cook all foods thoroughly, stirring and mixing to assure that heat is uniform throughout the food.

(g) Once foods have been prepared, do not allow to remain at room temperature for any length of time before serving. Temporary holding periods are 45°F. or below to 140°F. or above.

(h) Test temperature of food on steam tables to be sure it is not below 140°F.

(i) Heat leftovers thoroughly to center of food, do not merely warm. Never mix leftovers with fresh food to "extend" it. Remember that ground, minced foods are especially easy to contaminate because a large area of the food has been exposed during the grinding process. Keep leftovers for one day only.

(j) Correct washing and sanitizing of dishes, eating utensils and any equipment or surface which will contact food is absolutely essential to food protection.

CARELESS OPERATIONS AND LACK OF CLEANLINESS MAY LEAD TO DAMAGE CLAIMS AGAINST YOUR ESTABLISHMENT.

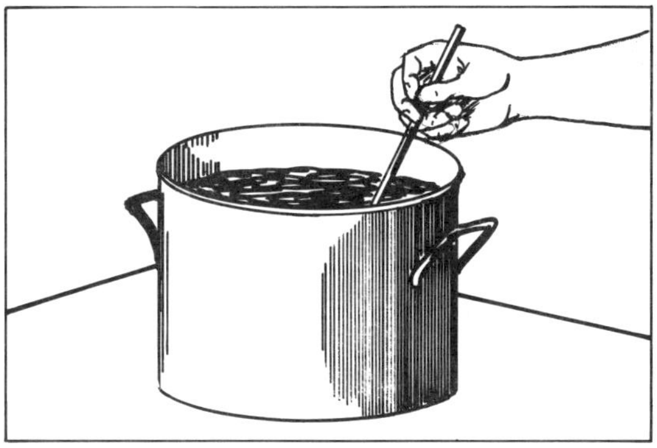

Large kettles or pots of such foods as chili, fish and meat mixtures, stews, soups or gravies should be stirred frequently while cooking or being held in steam tables. Temperature of all food in the kettle must reach 140°F. and remain there until food is served.

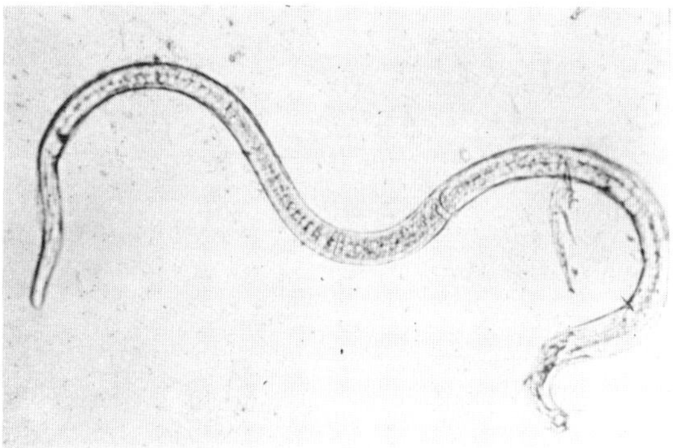

This is the trichinella worm magnified many times. To destroy these worms, pork must always be cooked thoroughly; must never be served red.

(k) Refrigerate food in small, shallow pans rather than in one large, deep container. Pans 3 or 4 inches in depth are suggested for storage purposes.

(l) Do not keep chemicals used to destroy pests in food preparation areas. This also applies to sanitizers, other cleaning materials. Do not wash scrub mops in dishwashing or pot sinks. All cleaning materials and cleaning tools should be stored in a cabinet used especially for that purpose.

(m) Hands and arms should be washed thoroughly before beginning work, after coughs or sneezes have contaminated the hands, after handling any soiled materials such as customer's used dishes and utensils, etc. Hands also need washing after cleaning poultry or handling raw meat.

(n) Protect food from droplet infection, flies, dust, etc. by covering it while it remains in the kitchen.

(o) Never put a spoon used for tasting food back into the food.

(p) Keep fingers away from mouth, lips, face and soiled surfaces. The hands should always be washed before preparing food.

Wrong: Never use hands to mix salads as pictured above left. Right: Always use spoon or fork to mix salads to reduce chances of contamination as illustrated at right. Serve salads with tongs, scoop or spoons.

Handle no food that can be transferred or picked up with tongs, scoops, forks, spoons, etc. Avoid mixing salads, ground meat, etc. with the hands. Never work with food if you have a sore, boil or infected cut on the hands or arms.

In several areas of the United States foodservice workers employed on cafeteria lines or in fast food service restaurants are wearing disposable plastic gloves while preparing or dispensing certain foods. The foods involved are those requiring no further processing by heat, such as cooked meats used in sandwiches; hamburger patties, wieners, bread, buns, cookies, vegetable sticks, fresh fruit, etc.

Users of the gloves consider them a valuable assist to sanitary food handling and a time-saver in food preparation and service. They act as a barrier between food and hands, thus food is not contacted by warm, moist fingers. Also, the wearing of the gloves is definitely appealing to the customer's esthetic sensibility.

It appears that disposable plastic gloves can be used satisfactorily in a foodservice operation provided certain rules are strictly followed. The mere wearing of the gloves does not guarantee a sanitary situation in the establishment, hence management and employees should not allow a false sense of security to cause any relaxation of required sanitation standards.

In the making of sandwiches, a disposable plastic glove can be worn on the hand which touches cooked meat. A glove should *not* be worn on the hand holding a knife or operating a slicer. Workers must thoroughly understand and carefully follow the safety rules pertaining to use of the slicer guard and the picking up of the cooked meat.

If the sandwich making is temporarily interrupted, the plastic glove should be discarded and a new glove put on when the work is resumed.

If worn for very long, plastic gloves—like hands—can become contaminated with various organisms. Hands wearing plastic gloves can brush against the hair, rub the eyes, scratch the nose, wipe on soiled towels, etc., if such gestures are habitual.

Sometimes foodservice workers neglect to report cuts or sores on the hands and may not cover such lesions properly. Here again, the plastic glove comes between hand and food, protecting food from contamination. But should the glove become torn or punctured while food is being handled, the food can easily become contaminated with a large number of staphlococci organisms. A similar situation of food contamination would exist if a punctured glove covers a perspiring hand.

It is suggested that disposable plastic gloves be discarded when they become obviously soiled; that, in fact, it would be safest to change them frequently during continuous work periods of preparing or dispensing foods. The gloves should be discarded when a worker's duties are interrupted for some other task, with fresh gloves being put on when the task is resumed; fresh gloves should be donned if those being worn are punctured or torn. Employees are cautioned to wash their hands after handling raw poultry or raw meat. Similarly, plastic gloves should be changed after the worker has handled raw poultry or meat and before he handles other foods.

A sanitation program involving the wearing of disposable plastic gloves will depend for its success upon the cooperation of management and the constant, vigilant supervision of those in charge of the employees using the gloves.

Keep foods off the floor. This is correct way to store dry foods.

(q) Store food off the floor.

(r) Don't smoke while preparing food. Saliva can be dropped on the food very easily while smoking. Saliva on the hands can contaminate food.

Milk in storage should be held at 40°F. to assure maintenance of best quality.

(s) Keep milk and milk products under refrigeration at all times.

(t) Milk containers, bottled drinks, fruit juices kept cold in ice should never be covered by melting ice water. If the ice water penetrates into the containers, the drinks can be contaminated.

Bread held in covered equipment is protected from insects and rodents.

(u) Preparation tables and work tables should be cleaned frequently with a germicidal detergent solution.

2. Sewage Contamination.

No overhead drainage pipes should be located over work tables, regrigerators or any place where unprotected food is stored. If such pipes are present, sheet metal drip pans should be suspended beneath them to carry condensation drippings and sewage drippings at the joints to a point of safety. No unprotected food should rest on the floor where it would be subject to contamination from any type of overflow or flooding. Sugar, flour or similar materials should never be stored on the floor, even briefly.

A refrigerator should drain through a trap to the open and then into the sanitary sewer. There must not be a direct connection to the sewer. This is to prevent the backflow of sewage due to any stoppage in the line and also to prevent the entrance of cockroaches, rats or sewer gasses.

Because drinking water may also become contaminated—especially in buildings where other floors are above the food establishment—local inspectors from the City Sanitary Engineer's office inspect the water piping arrangements to be sure no sewage or contaminated water could by syphoned into the food operation's drinking water supply.

3. Protection of food against rodents and insects.

All openings to the outside must be effectively protected against the entrance of rodents. Rodent-proof construction methods are described in *Rat-borne Disease Prevention and Control. V. Rat-Proofing of Existing Buildings; VII. Rat-Proof Construction of New Buildings.* This is a publication of the Public Health Service Center for Disease Control (CDC), Atlanta, Georgia.

Other insect and rodent control methods are described in the chapter on Pest Control.

Preservation Methods

1. Pasteurization

Pasteurization is the heating of foods, especially milk and milk products, to a temperature which destroys all harmful bacteria.

Since acid foods are generally capable of destroying bacteria at a lower temperature than non-acid foods, fruit juices, pickles, sauerkraut, etc. are often pasteurized rather than being subjected to higher temperatures.

2. Canning

Canning of food combines both cooking and sterilizing, with the foods to be preserved sealed in cans or jars. Pressure cooking is essential for the canning of those foods that may possibly be contaminated with botulinus bacteria, since the spores (outer protective coverings) of these bacteria can withstand high temperatures and are potentially dangerous.

Pressure cooking destroys all heat-resistant bacteria, provided the food is cooked for a sufficient length of time.

3. Cooking

As mentioned in previous chapters, cooking is the most important operation for destroying all harmful and most food-spoiling bacteria. Cooking, however, cannot be considered a sterilization process because some heat-loving bacteria are not destroyed by it. (These heat-resistant bacteria do not affect canned food since conditions are unfavorable for their development).

4. Dehydration

Dry heat is sometimes used to remove moisture from food in order to preserve it. In this case, the heat does not destroy the bacteria, but simply removes one of the conditions necessary for their growth. If dried food accidentally comes in contact with water during storage or packaging, the food quickly deteriorates due to bacterial activity. Foodservice workers should be careful to store all dehydrated foods in clean dry, covered containers.

5. Smoking

Meats and fish are preserved by being smoked for definite periods. The preserving action is due to drying rather than to the chemical preservatives (ex. creosote) which are found in smoke.

6. Refrigeration

Although refrigerators keep perishable foods from spoiling, food cannot be kept indefinitely in the refrigerator without showing some signs of bacterial activity. Milk sours; meat becomes slimy; butter becomes rancid. Refrigeration temperatures between 30°F. and 45°F. only check the growth of bacteria, they do not destroy them.

Most frozen foods can be kept satisfactorily for a year. At a temperature of, for example, minus 20°F. frozen meat, fruits and other products have been held for two years without any noticeable change of quality.

Slow and quick-frozen foods, however, are not sterile. Most of the bacteria are killed by very low temperatures, but some of them form spores and continue to live. During the thawing process, bacterial activity is resumed. If food is contaminated with bacteria which cause food-borne illness, harmful toxins may be produced during the thawing process. Therefore, the same sanitary methods employed in the handling of perishable foods must be used in the preparation of frozen foods.

The four main tasks of proper refrigeration are:

(1) To prevent the growth of dangerous germs.
(2) To preserve nutritive food value.
(3) To preserve food flavors.
(4) To save money by prevention of food spoilage.

Foodservice workers should immediately report any noticeable defect in the refrigerator's operation, such as door gaskets in poor repair or motors not operating efficiently. A rise in temperature above 45°F., as noted on the thermometer installed in the refrigerator, is a cause for great concern.

High humidity in the refrigerator aids temperature in penetrating foods and also keeps foods from drying. It must be controlled, however.

Food must be arranged in the refrigerator so as to permit the free circulation of air. Overcrowding, stacking foods one on top of the other retard the refrigerating process. Food in the refrigerator should be kept covered to prevent contamination from foods placed on shelves above, and to prevent condensation from dripping into foods.

It is suggested that, if raw foods and cooked foods must be stored in the same refrigerator, the cooked food should be stored on shelves above the raw food.

Refrigerators should be cleaned and sanitized regularly. This should be planned for a time when the least amount of food is in the refrigerator. All food should be removed during the cleaning and temporarily stored in another refrigerator. Refrigerator doors should be opened only when necessary. Frequent opening of the doors will cause the refrigerator temperature to rise, making the unit work harder and wasting power.

New purchases should be placed at the back of the refrigerator to ensure the use of older items before the newer ones. Leftovers should be kept no longer than 24 hours.

When food items are prepared in the establishment and frozen for future menu use, authorities in this field recommend that blast freezing be the method employed to produce quality frozen products.

A blast freezer, available in different models and freezing capacities, must be supplied and installed by experts. It is used to freeze products rapidly with blasts of sub zero air in minimum time to produce quality frozen products. The unit should be turned off when not in use.

The blast freezer must develop below zero degree air which is circulated with sufficient turbulence to maintain a chilling effect of minus 20° to minus 40°F.

When the temperature of the cabinet reaches approximately -20° to -40°F., the warm or refrigerated food items may be placed in it. Manufacturer's suggestions are that food products should be covered and sealed in conductive material such as shallow metal or aluminum pans. Preferably, the portions should not be over 2 in. thick and should be spaced on shelves for maximum air circulation.

When the product reaches 0 to minus 10°F., it should be removed and placed in a storage cabinet. Insulated gloves to protect the hands should be worn when working inside the cabinet.

A frozen food storage cabinet is a low temperature unit which will store frozen food products at approximately 0°F. Food packages should be rechecked before storing in the cabinet to be sure they are properly covered and sealed to prevent "freezer burn" or drying out of the products.

When bulk or convenience foods are moved from the frozen food storage cabinet to the refrigerator for defrosting, it may take from 24 to 48 hours for the items to thaw.

A unit is available which can be used at all times as a regular refrigerator with temperatures from 38° to 45°F. At the same time this unit has an accessory which is used to accelerate thawing of food products while maintaining a safe storage temperature not over 45°F.

The manufacturer's instructions for the cleaning and maintenance of this equipment are very clear and specific. These instructions should be placed in a convenient drawer or posted on the units for ready reference.

7. Chemicals

Chemical preservatives are sometimes added to foods to retard bacterial activity. They do not completely prevent the growth of bacteria, so foods which are preserved with chemicals remain sound for only a limited period of time. Many substances used as preservatives are in themselves rather harmless but have the ability to retard bacterial growth. Among these are sugar, salt, nitrates, vinegar, fruit acids, and alcohol.

8. Irradiation

Irradiation of foods is a fairly new method of preservation which will undoubtedly become more widely used when all its possibilities have been explored.

9. Ultraviolet Lamps

These lamps have been used to prevent the spoilage of meat and baked goods. The rays are helpful only to foods in good condition. If bacterial activity has advanced too far and the bacteria have penetrated into the interior portion of the food, the rays cannot reach them.

Protecting Convenience Foods

One of the most exciting developments in recent years has been the marketing and practically immediate acceptance by the commercial foodservice operator of convenience foods. More and more food processors are preparing frozen food items, some ready for consumption without further cooking, others which must be reheated or reconstituted in the establishment's kitchen.

Equipment manufacturers have cooperated by developing special equipment—steam cookers, revolving ovens, convection ovens, custom designed bain maries, reconstitutors, etc.—which meet the needs of foodservice operators who find that convenience foods permit more varied menus in less time and at lower over-all cost. Perhaps one piece of convenience equipment familiar to the largest number of foodservice employees is the microwave oven. This unit is used for fast reheating, defrosting and some primary cooking of small amounts of food.

From a sanitation standpoint, the microwave oven possesses several desirable features. We have learned that the length of time foods remain in the temperature incubation or danger zone (between 45°F. and 140°F.) has a direct relationship to the multiplication of any staphlococci present in the food and to the production of their toxin which causes illness. With the microwave method of cooking, the rapid heating of refrigerated foods reduces the "danger zone" time to the shortest possible period.

It has also been stated by authorities in the field of microwave energy that controlled experiments have demonstrated that short microwave heating cycles are as destructive to bacteria in the vegetative state as are the longer heat procedures. (This does not apply to organisms in the spore state which require higher temperatures and pressure to kill them.)

Many foods may be cooked in the serving dish or package in which they were purchased. If metal containers are to be used with safety in microwave ovens, the containers should be shallow and a reasonable food load must be in the container.

A steak, pre-seared under a broiler or on a grill, could be placed on a metal serving dish and finished to the desired degree of doneness in a microwave oven. A large rib roast could be cooked on a metal tray in a microwave oven with no problems.

Some users of microwave ovens prefer to use no metal containers or dishes with metal trim. The appearance of chinaware with metal trim can be damaged because the thin metal trim will tend to arc and pit. Oven-proof glass, ceramics, plastic ware, chinaware, paper containers of all types and plastic films are all transparent to microwave energy.

Pre-cooked frozen foods are often prepared in a material called polyethylene laminated to mylar. The food may be heated in this same package in microwave ovens, but a few punctures should be made in the pouch to relieve the pressure.

The principles and detailed practices of microwave cookery will be explained by the manufacturer's representative and by the very interesting reading matter he will leave with the manager or supervisor.

One fact to be remembered is a statement listed under the manufacturer's instructions for operation.

Make sure the microwave oven is clean at all times.

The same manufacturer's cleaning procedures follow:

To clean inside the microwave oven:

(1) With a damp cloth remove any spattering after each operation or as needed.

(2) Daily clean the entire cavity, shelf and door with warm water and mild soap. (The cavity is the interior area of the oven where the food is cooked.)

Do NOT use abrasives on stainless steel. Soak hard, dry particles with hot water.

To clean outside of microwave oven:

(1) Wash outside with warm water and mild soap. Rinse and dry.

(2) Clean the oven's plastic window. First, loosen the screw retaining the window. Unlatch the door handle and swing the hinged plastic window away from door face.

Clean both sides of window with warm soapy water. Rinse and dry. *Do not* use hard abrasives or detergents.

With door handle in unlatched position, close window and tighten screw. Close door.

(3) Clean filter under front of unit, depending on use, oven placement and dust particles in air, with mild soap and warm water.

(4) Keep control panel clean.

(5) To remove fingerprints, apply a small amount of mineral oil with a clean cloth, rubbing with the grain of the steel.

It is pointed out that if the oven is installed too near a broiler or griddle, the possible grease accumulation on the filter will necessitate more frequent cleaning.

From the standpoint of food protection, health authorities point out that the processing of these food items should be carried on with the same strict adherence to sanitation standards accorded any other processed food. Equipment manufacturers stress that their instructions for reheating and reconstituting should be followed exactly. It remains, therefore, for the foodservice operator to treat convenience foods, insofar as storage, thawing and service are concerned, as any other highly perishable frozen foods are handled in the establishment.

Convenience foods are becoming more varied and popular. The combined efforts of the processor, the equipment manufacturer and the foodservice operator should prevent, as far as possible, any occurrences of food-borne disease due to consumption of these food items.

SUMMARY

Food is a major investment, both to the commercial food operator and to the homemaker, to say nothing of its cost to hospitals, universities and other types of institutions, school lunch programs, etc. In addition, its preparation and service provide employment to thousands of Americans.

The most recent available figures show that 169 million meals are served daily to Americans who eat out. On an annual basis, the figure amounts to 55,770,000,000. This includes all types of institutions as well as commercial establishments and takes into account the closing of school lunch programs for a part of the year, the one-day-a-week closure of many restaurants, etc. These are astounding figures.

It follows, then, that food must be protected by handling according to sanitation regulations from its very source to the moment it is placed before the customer. When all those connected with the foodservice industry recognize this fact and apply its meaning to their own phases of foodservice activity, we should have no more cause for concern about food-borne disease outbreaks.

FOOD PROTECTION QUIZ

1. True or False (check True or False): True False
   (a) Food spoilage is caused by bacterial action. ( ) ( )
   (b) Spoiled food cannot be detected until it has been eaten. ( ) ( )
   (c) Food-borne illness and food spoilage are caused by the same germs. ( ) ( )

2. Fill in the following statement:
   (a) In foodservice operations, food considered unfit for use should be ____________.
   (b) "Sick" food has been contaminated with ____________________bacteria.

3. Check correct statements.
   (a) Time and temperature are two important factors in food protection. ( )
   (b) Never disturb food by stirring it while it is cooking ( )
   (c) Taste soup often with a ladle to be sure of proper seasoning. To save time, use the same ladle. ( )

4. Fill in the following statement:
   Temperature in center of cooked pork should be ^{o}F. or higher.

5. True or False (check True or False): True False
   (a) Keep leftovers until they have been used up. This saves money. ( ) ( )
   (b) Never mix leftovers with fresh foods. ( ) ( )
   (c) After 24 hours, throw away leftovers. ( ) ( )
   (d) Convenience foods are so named because they require no special care. ( ) ( )

6. Check correct statements:
   (a) Proper washing and sanitizing of utensils and equipment are necessary to food protection. ( )
   (b) There is nothing wrong with smoking while preparing food. ( )
   (c) Washing the hands thoroughly before food preparation helps to protect the food. ( )
   (d) Pasteurization is a type of cooking. ( )

7. Fill in the following statements:
   (a) The safest method of canning foods is to use a ____________________.
   (b) Dehydration preserves food by ___________ its ________________.
   (c) Name two common foods often preserved by smoking. ______________,__________.

8. Can food be kept indefinitely in the refrigerator without damage? Yes ( ) No ( )

9. How many meals per day do Americans eat out? Check correct answer.
   800,000 ( ) 169 million ( ) 1 billion ( )

10. True or False (check True or False): True False
   (a) Use meats and poultry only from approved sources. ( ) ( )
   (b) Pour milk for customer into a glass from large container before serving his meal. ( ) ( )
   (c) If you get too warm while working at the steam table, turn down the heat. ( ) ( )
   (d) Never handle food in your hands if the task can be done with a proper utensil. ( ) ( )

6 Pest Control

For many centuries man has been waging war against those harmful insects and rodents which destroy his crops, eat and contaminate his foodstuffs, spread various diseases, and disturb his economy. Though some of his battles have been won with spectacular success, other important and very expensive campaigns are still being interfered with by traitors within man's own ranks. These are the many unthinking individuals who assist the enemy by providing living quarters and food for those insect and rodent pests so closely associated with mankind they are now termed as domesticated.

Included among the relatively few types directly involved in the cost and sanitation phases of foodservice operations are, unfortunately, some of the smartest examples of insect and rodent life. Research workers, for instance, have proved that the reaction time of the cockroach is five times as fast as that of the trained athlete. The intelligence of the rat has long been recognized.

These pests can be controlled by modern sanitary practices, backed up by proper chemical treatment. Foodservice workers play an important part in the control picture.

INSECTS

Of about 1,000,000 varieties of insects thus far identified by scientists, only a few thousand have bothered man enough to make him interested in doing something about it. And of these few thousand the real problems are due to about 200 varieties.

Although many of these are pests in the food industry as a whole, our discussion will be limited to those insects most often found in the foodservice establishment.

1. The House Fly

The house fly is one of the most common domestic insect pests. It is also one of the dirtiest. People (including foodservice workers) who tolerate house flies and brush them off food with apparent unconcern undoubtedly do so because they are not familiar with this insect's breeding and living habits.

The fly breeds in filth such as garbage, human and animal wastes, rotting vegetative matter, etc. Mankind also provides him with enjoyable living conditions such as: saliva on sidewalks, open garbage cans, dumps, public toilets, unprotected and spilled foods. DEVELOPMENT OF THE FLY—The female fly lays masses of eggs in filthy material. In a few hours these eggs hatch into maggots, the white, wormlike, soft-bodied larvae of the insect. The maggots burrow into the breeding material, where they remain and feed for periods extending from a few days to several weeks. After the feeding state, the larvae look for some dry, sheltered place and enter the next period of development, the pupa stage.

Activity ceases for from 3 to 10 days while the larva inside the pupal case is changing into the flying insect. When the change has been completed, the fly pushes out of its pupal case, makes its way to the surface, spreads its wings and waits for its body to dry and harden. It is now ready to begin an existence of 3 or 4 weeks as an adult fly when the female may deposit 2800 eggs.

The development of the fly from egg to adult has been described briefly to show

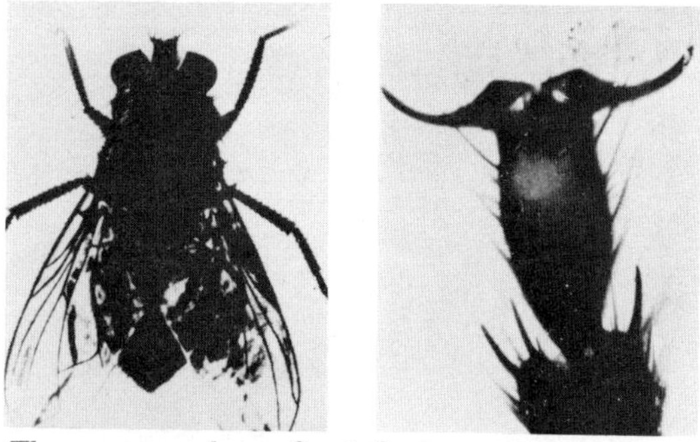

The common housefly, left above, is a menace to safe food. The enlargement at right of the foot of a housefly stresses the dangers that exist when flies are permitted to move freely about a foodservice establishment.

that the circumstances of its breeding period cause it to be unclean from the first moment of its existence.

THE BODY OF THE FLY—Many high school science students have examined a fly under a powerful magnifying glass. They have seen a very hairy body, with feet which resemble delicate transparent webbed leaves. Two hairy, flattened pads on each foot allow a sticky material to ooze out. It is this sticky substance which enables the fly to walk on ceilings, or cling to thin objects like wires, hanging cords, or strings. This same substance collects filth and germs which are later rubbed off on food or any other surface the fly happens to contact.

The fly has no teeth. He does possess a sort of snout with a spongy mouth part at the end. This organ is necessary to his peculiar feeding method.

HABITS OF THE FLY

(a) Eating Habits—when the fly eats, he sucks liquids into his spongy mouth part. He also likes solid foods, particularly sweets like sugar, candy, cake frosting, etc. In order to eat solids, he regurgitates (brings up) a fluid from his stomach. This passes through his snout and is deposited on food to soften and moisten it, thus enabling him to suck the nourishment back into his body.

The fly often eats and eliminates at the same time. The light colored spots seen on feeding areas are the regurgitated stomach fluids; the darker spots are eliminated wastes. Both are called fly specks.

These fly specks are the visible signs of a fly's presence. They should remind foodservice workers that the fly visits many different areas in search of food, hence could have come to unprotected food direct from garbage or filthy material. Its body and feet have picked up filth which could well contain disease germs. Some of these germs are present in the regurgitated fluid and body wastes. This is another way the fly can spread disease-causing organisms in a food establishment, in addition to rubbing them off from his feet.

A New York State report says that laboratory examinations of the outside of a fly's body and its alimentary canal revealed as many as 28,000,000 bacteria on the inside and 5,000,000 on the hairy body and feet of a single fly.

(b) Living habits—the fly's sense of smell attracts him to such items as: uncovered garbage, filthy toilets, spilled food, soiled paper napkins, meat scraps, soppy straws, empty milk cartons, etc. If flies are seen in any number around food establishments, it can be taken for granted that they are breeding and living in the close neighborhood. They live and breed no farther away than 300 to 500 yards from the place where they are seen. They may have a flying range of 13 to as much as 25 miles and may be transported from place to place in automobiles, planes, trains, or by riding through revolving doors on the clothes of customers: but if the number of flies is a problem, look for breeding places around the building. Look for damp areas on the ground or in refuse and if maggots are detected, the establishment is helping to breed most of its own flies.

(c) Diseases spread by flies: The fly is both filthy and dangerous because of the diseases he may spread. He is thought to play an important part in the transmission of many disease germs, especially those which are found in the intestinal tract or the nose and throat of man. Some of the most important of these are tuberculosis, typhoid fever, dysentery and salmonellosis.

FLY CONTROL METHODS—Food operation fly control includes: elimination of breeding places; elimination of fly-attracting odors; keeping flies out of buildings; destroying flies which gain entrance; proper food storage and food display; proper uten-

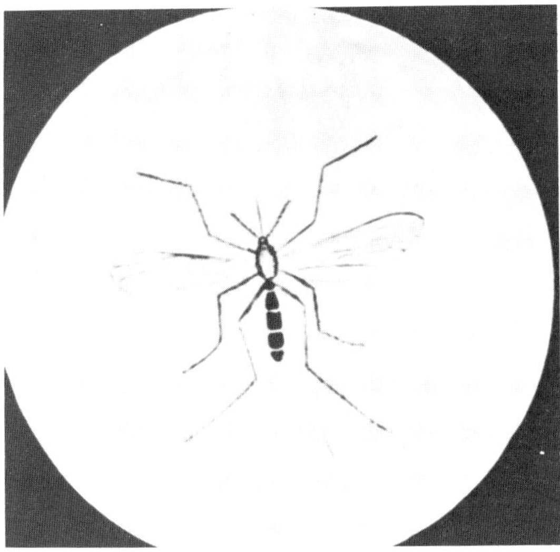

Even the mosquito can be considered a contaminating pest. Aside from the annoyance to customers, an infected anaphelis mosquito may bite a patron or a malaria sufferer may be bitten by this mosquito. This is of particular concern in those areas where malaria is still a problem.

sil storage; killing back door populations.

A number of sprays, both space and surface as well as aerosols, are available for the killing of flies. Fly strips are also marketed.

Consult the health department to be sure an insecticide has been approved for use in commercial food establishments in areas where foods are prepared, served or stored.

Fly strips, for example, may contain a chemical not recommended for use in any area where foods are prepared, served, and/or stored. Some health agencies rule against the use of particular strips in hospitals. nursing homes for the elderly, or nurseries.

Strips containing the same chemical may be used effectively in non-food areas, such as the inside of a garbage can lid; in a completely separate, enclosed garbage room; or in other locations removed from food.

Labels should always be carefully checked for use purposes. Even when this is done, the health department should be the final authority. Sometimes after investigation and research into new products–such research often lasting for several months–regulations may be relaxed in some areas to permit more extensive use of the products under very carefully controlled conditions.

The pesticide company's field representative should be consulted as to the safety and advisability of using insecticide products in the establishment during the intervals between his regular visits. If personnel *are* selected to use such products, they should be carefully instructed and supervised while working with the product.

In conclusion, remember that the fly may spread disease germs in three ways:

By its body wastes
By the way it eats
By its feet–it leaves its tracks

Always remember that the practice of sanitation, the elimination of breeding areas and the assistance and advice of a qualified pesticide operator, if necessary, are the factors contributing to a successful pest control program.

2. Ants

These insects are called social because they live and work together in social groups or colonies. They are pests from an economic standpoint rather than disease-causing, but their continued presence in a food-service operation is an indication of poor sanitation. Some ants are fond of sweets and fatty foods, others prefer meats and grease.

Nests are often large, contain thousands of ants and are found between walls, under floors, above ceilings, behind baseboards, in old trash or outside the building. ANT CONTROL METHODS–Practice sanitary measures in the establishment, keep foods protected, spray around baseboards of building, spray outside of buildings and in infested areas (with special attention given to ant trails).

If personnel are assigned to control operations. consult pesticide company's representative for suitable sprays and advice on how to apply them.

3. Cockroaches

The cockroach, another filthy contaminator of food, is a major insect pest. Five varieties of cockroach are commonly found in the United States–the American, oriental, German, Australian and brown-banded.

The American cockroach, the largest variety, is about 1½ inches in length, red-

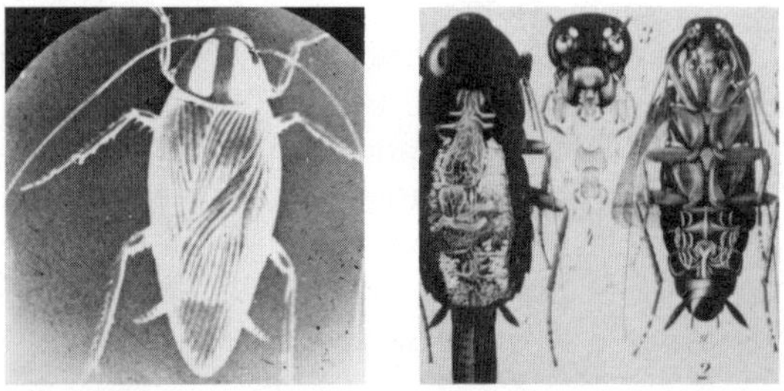

The cockroach, left above, is another contaminator of food. Cross sections, right above, are of male and female cockroaches. The capsule of young cockroaches present in the female can be noted as an indication of the tremendous reproduction rate among roaches.

dish brown in color, has light markings on the wings, legs and breathing vents, and well developed wings. It prefers warm, damp places like cellars, but may also be found in cartons and shipping cases.

The oriental cockroach is often called a water bug. It is dark brown, about 1 inch long; males have fully developed wings.

The German roach is smaller–about ½ inch long–moves very fast, rarely flies, and can be found in almost every area of the United States. It likes to live in kitchens.

The Australian roach, somewhat smaller than the American variety, prefers vegetables.

The brown-banded roach is much like the German variety except that it flies. Its body varies from black to a light golden color. It likes cupboards and shelves.

The female roach drops or attaches egg capsules from which the young will hatch in such places as: behind store windows, under shelves and counters, behind furniture, in folds of drapes, behind and under drawers, behind and under food equipment, in cracks and crevices.

When a food establishment has a large roach population, they may be found literally everywhere, from elevator shafts to washrooms.

Unless the establishment is heavily infested, the roach may be seen only at night. It is sensitive to light and prefers to live and breed in the dark. Foodservice workers assigned to use aerosols to drive these insects out of hiding places can state that they will swarm out of foodservice equipment, cracks and crevices, from behind baseboards and sinks, even from meter housing.

Roaches do not have to be seen to make their presence known. Discarded egg capsules and body wastes are sure traces that they are living in a foodservice operation. In addition, some cockroaches, while eating, regurgitate a brownish fluid which has a very foul odor.

Travelling in the variety of places they do, from sewers to garbage to food, cockroaches feed on contaminated materials and pick up large numbers of germs. These germs may be present on the roach's body or be taken into its stomach, where they may remain alive for a long time. Thus, they have been implicated in carrying the organisms causing diarrhea, dysentery, typhoid fever and food poisoning to unprotected food.

COCKROACH CONTROL METHODS

(a) Good building construction. (Wherever possible, fill cracks and crevices with suitable materials. Seal pipe openings. If doors must be propped open for receiving produce, etc., be sure they are closed as soon as possible. Proper lighting and ventilation are disliked by any insect preferring dark, humid conditions for breeding purposes and harborages.)

(b) Good housekeeping.

(c) Storage of food in covered containers.

(d) Use of approved insecticides.

(e) Carefully inspect all produce that comes into the establishment. Roaches may often be found in bags or cartons.

RODENTS

1. Living Habits

The rat is one of the most dangerous and costly pests with which man must contend. It is believed he first reached the United States on ships from foreign lands.

The Norway rat and the roof rat are the two most common species in this country today. The Norway rat is also called the sewer, brown, barn or wharf rat. The adult weighs about a pound and grows to a length of 10 inches with a 6 inch tail. His fur is a coarse reddish-brown but may appear grayish when coated with dirt.

The female averages 4 to 7 litters of

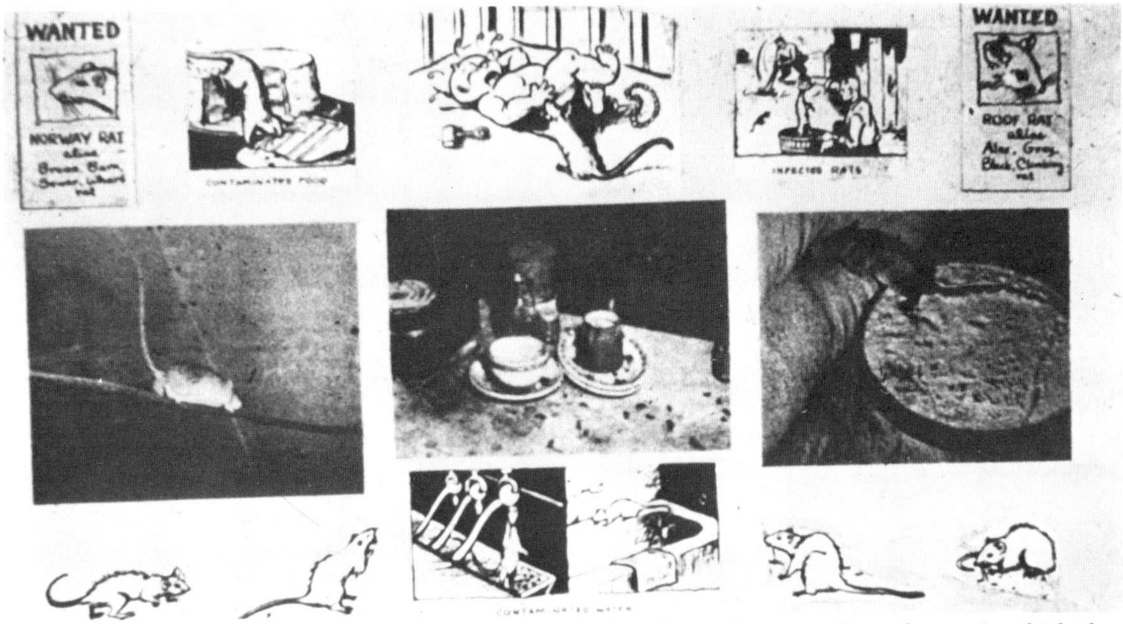

The rat is one of the most dangerous and costly pests man has to put up with. Dramatic illustration above shows number of ways in which they carry diseases.

about 10 young per litter per year.

The Norway rat is a strong swimmer but a poor climber. Hence his runways in establishments will be most often found along the base of walls. He often uses city sewers for runways. He will eat either fresh or decayed food and will travel from one building to another to find what he wants. He will gnaw his way through wooden doors, soft cement, boxes, etc.—any material less hard than the enamel on his teeth. It is estimated that he requires one ounce of dry food and onc ounce of water daily.

The roof rat is an excellent climber. He prefers warm climates and enjoys vegetables, meats, fruits and grain.

This "rat-run" shows dirt and grease marks from rats' bodies, present because rats use this same path over and over again to gain access to building. Rat-proofing is only positive means of controlling entry of rats.

The common house mouse is a dusty gray animal weighing about 1½ ounces. His tail is as long as his head and thin body. Mice mature in only 45 days, live about a year, with the female averaging 8 litters of 5 or 6 young each year. Mice are food nibblers rather than hearty eaters.

Rodent control should be a community project rather than an individual establishment program. The business sections of many cities may be honeycombed with rat runs, making the elimination of the rodent from scattered foodservice establishments in the area a difficult undertaking on a long-lasting basis.

REMOVAL OF THE RAT'S FOOD SUP-

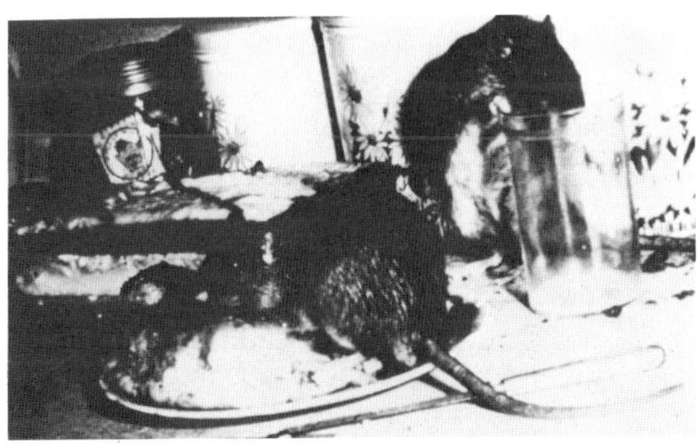

An alarming example of what could happen when food is left to stand uncovered overnight if rats are not controlled.

PLY—Keep garbage in closed metal containers with tight fitting lids. An efficient system of garbage collection and disposal is necessary for the control of rats in all types of foodservice operations.

Compactors as a solution to the trash disposal problem are becoming increasingly popular in food establishments of all types. Various models and sizes are available for restaurants, motels, hotels, nursing homes, hospitals, fast food establishments, bars, offices—practically all types of institutional operations.

The compactors can be used as a major means of waste control or as a support or auxiliary aid, no matter what the method of final waste disposal may be.

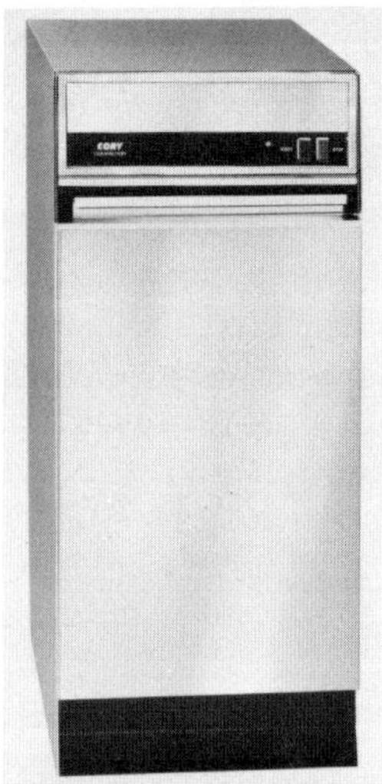

This compactor is sized for foodservice use in quick service operations, nursing homes and similar institutions. Also meets needs in bars.

Variously sized models have been designed for foodservice establishments of all types. One small size unit of counter height will compact glass, cans, boxes, light bulbs, bottles, cartons, food wastes, etc. to a total of 35 lb. before unloading is necessary.

Another model has a system whereby the time and labor required to transfer boxes of compacted waste to outdoor storage areas is eliminated by installation of a passage through the wall.

If correctly used, compactors would seem to be a definite aid to the sanitation program by helping to eliminate problems due to insects or rodent infestation.

STORAGE

Store all food so rats cannot get to it. Rodent-proof storerooms may be constructed out of any material such as ¼ inch hardware cloth, which will resist the gnawing of rodents. No hole or crack should be permitted in the construction. Store food at least 8 inches above the floor, better if it is 12 inches. Aisles should be wide enough to permit inspection of stored food, including the area around the walls. No food should be left exposed overnight.

REMOVAL OF HARBORAGES AND NESTING PLACES—Since rats build their nests in rubbish heaps, under piles of lumber, in hollow spaces of poorly constructed walls, in cellars and other places previously mentioned, all such conditions should be eliminated.

Rats are tricky so to trap them be sure to place trap correctly, in areas where rats are likely to travel—runways, along walls, in cellars. Attach bait securely.

TRAPPING—The spring trap is easily the most effective.

Attach bait to trigger, then place traps where rats are apt to travel, such as runways along walls, or behind and between objects. Rat runways may be easily detected. They are the regular routes taken by rats in search of food and are stained with dirt and soil from its body. Numerous rat droppings are another indication of areas often used by rats.

The traps should be placed at right angles to the runway.

POISONING—Use great care in handling rat poisons. There must be no danger of contamination of foodstuffs or personnel.

SUMMARY

1. **Keep rats out!**

   Repair—Breaks in screens, windows,

doors, etc. should be immediately repaired.

Block out—The openings around wiring, piping, fans, vents and flues should be repaired. A rat can squeeze through any place a clothes pin can be inserted.

2. **Sanitation.**

Operations must be spic and span—in other words, good housekeeping.

Garbage disposal should be proper and adequate. Garbage rooms should be cleaned and disinfected regularly, as should garbage cans.

Keep premises both inside and out really clean. This means a continuous sanitation program.

When rats are numerous in an area, two conditions can always be found:

(a) A food supply is easily available.

(b) There are places for rats to harbor and raise their young. It has been said that there are usually 20 rats for every one seen on or around the premises.

DISEASES TRANSMITTED THROUGH RATS

(a) Food Poisoning and Food Infection

These may be caused by contamination of food with waste matter from either rats or mice.

(b) Salmonella organisms are found in the intestinal tracts of rats and mice and can be deposited on food.

(c) Endemic Typhus Fever.

This disease has its source in the infected rat and is transmitted by the rat flea when the flea bites a human being.

(d) Bubonic Plague.

Plague is a disease often associated only with the Dark Ages; infrequent cases have been reported in the United States, brought in, probably, by infected visitors who arrive in our country via ship. Here again, the rat flea is the transmitting agent.

(e) Ratbite Fever.

The method of transmission is from the bite of an infected rat.

(f) Trichinosis.

As scavengers themselves, rats eat infested flesh, become infested and are in turn eaten by scavenger animals such as pigs.

RODENT CONTROL METHODS

Rat control measures for foodservice establishments are similar to those required for community wide control and include:

(a) Ratproofing of buildings.

Ratproofing is the best way to ensure a permanent control.

Many communities have adopted ratproofing ordinances and provide for a continuing control program.

Architects, plumbers, building organizations and contractors are, more and more, using the principle of ratproofing in new construction.

Examples of ratproofing in a building include metal flashings on doors and metal collars around pipes; metal screening over basement windows, and the installation of door-checks to insure doors closing tightly. Advice and assistance will be provided by local health departments.

(b) Use of chemicals

Insecticides and rodenticides are not a substitute for sanitation, but rather a supplement to the sanitation program. Clean surroundings are appealing to man but disliked by pests.

The most efficient way to eliminate pests on a temporary basis is by the employment of a professional pest control operator. "Temporary" is used here, because the control efforts of the most efficient operator cannot be considered on any other basis except temporary without the cooperation of the establishment in eliminating the pests' food and breeding places.

As time passes, both insects and rodents become resistant to previously used chemicals and new ones must be developed. The professional pest control operator knows what to use and where and how to use it. He will also train personnel in proper procedures, then visit the establishment to inspect and supervise the control program.

Finally—Follow the rules of sanitation outlined in these chapters and BE ETERNALLY VIGILANT.

PEST CONTROL QUIZ

1. Check statement or statements below that are correct.
   (a) The house fly breeds in the water. ( )
   (b) The house fly breeds in filth. ( )
   (c) The house fly breeds in a nest. ( )
   (d) The house fly breeds in human wastes. ( )

2. Fill in blanks in sentences that follow:
   The house fly has a ________________ body. It can walk on the ceiling because of a ________________ which oozes out of pads on its ________________.

3. True or False (check True or False):

The fly takes in liquids by:	True	False
(a) drinking them	( )	( )
(b) sipping them	( )	( )
(c) sucking them up through its snout	( )	( )

4. Four diseases possibly spread by the house fly are:
   1. ______________________________ 2. ________________________
   3. ______________________________ 4. ________________________

5. Fill in the following statement:
   Flies may spread disease by ________________ or ________________________ or ________________.

6. Do ants spread food-borne diseases? Yes ( ) No ( )

7. True or False (check True or False):

When ants are always around in a food establishment it is because:	True	False
(a) They like to be around people	( )	( )
(b) There are poor sanitation practices in the establishment	( )	( )
(c) Ants prefer to live indoors	( )	( )

8. Check 4 diseases in which cockroaches have been implicated as transmitting agents.
   dysentery ( )
   mumps ( )
   diarrhea ( )
   acne ( )
   typhoid fever ( )
   food poisoning ( )

9. Check two conditions which rats look for in desirable living quarters.
   ( ) places to hide; ( ) warm temperature; ( ) an easily available food supply.
   ( ) access to sewers; ( ) ropes to climb; ( ) a suitable nesting place to raise their young.

10. Fill in the following statement:
    Efficient rat control requires the cooperation of ________________________ and the ____________________________ operator.

The Teaching of Dishroom Procedures

Notes for Instructors:

The problem of keeping the entire group alert and attentive during presentation of the washing and sanitizing material is one which arises occasionally to plague instructors. When it does occur, some instructors question both their own ability and the value of including these lectures in sessions attended by all establishment personnel. Why not leave it for the ears of the dishwashing crew alone?

There is usually one main reason for inattention and restlessness. A class consisting of personnel from all departments of the establishment almost always includes a few who question their need for this particular instruction as well as its practical value as an aid to their own job efficiency. Since their duties are far removed from the dishwashing area, they see no reason to listen closely and their obvious disinterest can easily distract other class members.

Such an attitude should be recognized and corrected before the instruction begins. Point out that:

1. Every employee must be concerned with the work of the dishroom crew. If tableware is recontaminated after proper washing and sanitizing, management's expensive investment in dishmachine equipment and detergents has been wasted, the sanitary procedures practiced by a well trained dishroom crew have been of no value, and the employees who have mishandled the tableware can rightfully be termed either unaware of the necessity to follow sanitary procedures in dish and utensil handling—or they are unforgivably careless.

2. If, on the other hand, the dishwashing crew is at fault so that food is served on unclean tableware, who bears the brunt of the customer's displeasure? It will be a waiter, waitress or other employee who has had no actual part in the cleaning process.

Dish and utensil sanitation is one more example of a foodservice activity involving the entire employee team. The carelessness of one will nullify the conscientious efforts of all the others.

Tell the class they will be rated equally on their ability to answer questions, join in discussions and offer suggestions. The dishwashers' interest will be stimulated—after all, they are supposed to know more about these subjects than the others; the remainder of the class will realize they have learned a great deal about one of the establishment's most complex operations. They will certainly understand that washing and sanitizing are not the dull, drab series of boring repetitions they had previously thought them to be.

Emphasis must, of course, be placed on the dishroom personnel's instruction in and understanding of every detail of their job routines. This must be followed up by competent supervision.

Meanwhile, do not expect your dishwashing crew to become chemists, experts on detergency or equipment engineers. They should, however, be well enough informed to know what to expect from a properly functioning dishmachine. Whether they wash by machine or by hand, they should have a working knowledge of the types of food soil to be removed and an understanding of why various cleaning agents, in strictly specified amounts, have been provided for the satisfactory removal of these soils. Such knowledge will often enable them to detect and report in advance on malfunctions which might otherwise result in major problems often leading to unnecessary management expense, loss of time and reduction in operational efficiency.

The dishwashing crew is entrusted with the day-by-day operation and care of one of the most expensive pieces of equipment in the establishment. Too often they are literally turned loose with less training or knowledge than it takes to punch the time clock. Management's investment in dishwashing equipment can be properly protected only when the dishroom crew is as thoroughly trained and as well supervised as any other group of employees in the establishment.

7 Manual Dishwashing

You may encounter some foodservice installations where the size and nature of the operation make washing dishes and silverware by hand the feasible method. Satisfactory results may be achieved provided a few simple rules are conscientiously followed.

Crews accustomed to machine use should also become familiar with manual procedures. In the event of mechanical breakdown during business hours, tableware would probably be washed and sanitized in the sinks used for pot and pan washing. If this area is properly equipped, such temporary emergencies can be met with little, if any, loss of sanitation effectiveness.

In manual dishwashing, the first requisite is adequate space in which to operate. The U.S. Public Health Service Food Service Sanitation Ordinance and Code, upon which many State and local food regulations are based, requires the provision and use of a 3-compartment sink wherever washing and sanitization of equipment or utensils are conducted manually. The few exceptions to this rule need not concern our present discussion of routine operations.

Sinks must be of adequate length, width, and depth to permit complete immersion of tableware, equipment or utensils and each of the 3 compartments should be supplied with hot and cold running water.

Dish tables and drainboards must be spacious enough to permit proper handling of soiled tableware prior to washing, and clean tableware after washing and sanitization. Construction of sinks, dish tables and drainboards should be of galvanized metal or better, should resist denting and buckling and be sloped so as to be self-draining.

All brushes and other tools, dish baskets, etc. designed for use in the dishwashing operation should be readily available, kept clean, and replaced whenever necessary.

Prior to washing, dishes should be sorted and stacked on the soiled dish table or drainboard. Stained dishes or those with dried-on or baked-on food soil will need soaking. A large container holding a warm detergent solution may be placed on the floor under the drainboard and conveniently used for this purpose. Special brightening and stain removing detergents are available if necessary. Your detergent field technician can advise on their use.

Silverware is also soaked to loosen food soil and tarnish. Usually a shallow aluminum pan or one having the bottom lined with aluminum foil is filled about half way with a warm detergent solution. Again, there are special silver dips and special soaking procedures developed by the leading detergent manufacturers, for use when indicated.

Washing Operation

The washing operation proceeds as follows:

1. Scrap, pre-rinse or both.

Purpose: To remove loose soil from dishes, thus keeping wash water cleaner, freer of bacterial population build-up. The cleaner the wash water, the easier it is to maintain detergent strength at required levels:

Method: Scrape gross food soil by hand or squeegee into table opening to garbage container beneath soiled dish table. Pre-flush loose soil from dishes with power spray attached to flexible hose arrangement installed on soiled dish table at drain provided for this purpose.

2. Wash in Sink No. 1.

Purpose: To remove all visible soil from dish surfaces, including the backs of dishes.

Method: Fill sink to proper level with clean, hot water (120°-125°F.). To ensure correct water temperature, have permanent thermometer attached to wall with bulb (protected) in wash water or keep a thermometer available in safe place for frequent testing of water temperatures.

Add specified amount of a detergent especially compounded for hand dishwashing.

(a) To measure the detergent accurately, it is necessary to know how much water is being used in the wash sink. With a 5-gal. can, fill the wash sink to the level usually employed. Determine the number of gallons added, make a permanent mark at the water level and fill the sink to this mark for every washing operation. When detergents are dispensed by hand, the usual instructions specify a certain amount per gallon of water. Knowing the amount of water in the sink allows the dishwasher to measure the detergent with reasonable accuracy. A plastic measuring glass for the detergent is easily available.

(b) Methods of feeding liquid detergents directly from the shipping container, without manual transfer of any kind, have been developed by leading detergent manufacturers.

Various uncomplicated devices are available which, attached to a single or combination faucet, will release pre-determined, measured amounts of detergent into the washwater. Excellent results can be achieved with these modern tools, always provided they are installed, used and maintained in strict accordance with manufacturer's instructions.

(c) For washing, use a suitable brush with bristles firm enough to provide the required friction. Clean and store brushes properly, replace as soon as they become too worn to clean effectively. Friction is an important part of the cleaning process, necessary to thorough soil removal.

3. Rinse in Sink No. 2.

Purpose: To remove washing solution or other material from dish surfaces.

Method: Rinse thoroughly in warm water, 100°F., until dishes are clean to sight and touch. For effective results, the rinse water must be kept clean. Let it run continuously and overflow.

Remember that all wash and rinse procedures lead to the final step in Sink No. 3, and bactericidal treatment of the tableware. And since the usually employed methods of sanitizing do not act effectively in the presence of soil on equipment or utensils, it is not likely that unclean equipment will be adequately sanitized.

After rinsing, dishes and silverware should be placed in suitable baskets or containers prior to sanitization. The principal reason for this (aside from the fact that we may be using hot water at too high a temperature to be borne by the hands) is to reduce hand contact with food surfaces of dishes and "business" ends of silverware to an absolute minimum. Thus we prevent recontamination and the waste of all work accomplished in the cleaning operation up to this point.

4. Sanitize in Sink No. 3.

Sanitizing and the approved methods of handling, storage and re-use of sanitized dishes and utensils will be covered in detail in a later section. We have correctly scrapped, washed and rinsed our tableware. We have taken the necessary steps to ensure its proper sanitization. You must recall these steps when we resume our discussion in the section on sanitizing.

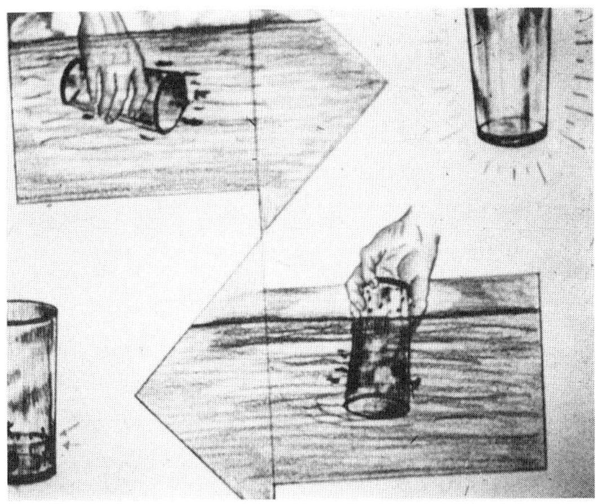

Correct and incorrect way to immerse a glass in rinsing or sanitizing solution.

Handwashing of Glasses

It is suggested that glasses be washed first, when the washing solution is cleanest. Use the type of brush or a sponge which will reach all inner glass surfaces. Often used are stationary brush sets which adhere by suction to the bottom of the sink. Glasses of intricate design or unusual shape will require special care to remove all soil. As with dishes, the object is to thoroughly cleanse during washing and completely remove washing solution or other material during rinsing.

Handwashing of Silverware

In addition to pre-soaking, silverware also requires a clean, hot detergent solution for washing plus the application of friction to remove all soil. Sharp inspection is necessary during the rinse operation. If any soil remains on the tines of forks, etc., the utensil should be resoaked and rewashed.

It is not possible to wash and/or sanitize effectively utensils or equipment with broken places, cracks, or open seams. Such places provide places for food to collect and sites where bacteria can grow.

Cracked Dishes

It is impossible to effectively wash and/or sanitize dishes or equipment with broken places, cracks or open seams. Food particles collect in such areas and growing sites for bacteria are provided. The only safe solution is the immediate discarding of any tableware with such defects.

Cleaning Pots and Pans

The same preferred procedures of scrap, pre-rinse, wash, rinse and sanitize apply to both the handwashing of tableware and pots and pans. After preliminary scrapping, pots and pans are often left soaking in a warm detergent solution in Sink No. 1. While this is an excellent practice, trouble develops if the same solution, grown cold and greasy, is used for the actual washing operation. The soak solution should be drained, the sink cleaned, and a fresh, hot detergent-water solution used for the pot and pan washing. Here again, the proper brushes should be made available.

In many foodservice operations, mechanical dishwashing is employed for glasses, dishes and silverware, but manual procedures are applied to pot and pan washing. One supplier of foodservice equipment suggests that cooking utensils be carefully hand washed, especially for removal of cooked-on food soil. Then they should be put through the dishmachine, if size permits, for rewash and final sanitization.

Management may provide a liquid detergent, perhaps one drawn directly from its shipping container to the sink and measured by one of the previously mentioned faucet fixtures, or a powdered cleaner or specially made bar may be employed. The important thing for pot and pan washers and dishroom supervisors to remember is that no detergent will work effectively and no amount of physical labor will clean effectively when the wash solution has become cold, full of grease and food residues.

A clean, overflowing rinse, just as necessary for pots and pans as for dishes, should be provided in Sink No. 2.

The sanitizing procedure in Sink No. 3

Three compartment sink with circulator for soak compartment. Also can be used for pot and pan washing.

will be discussed in the section on sanitizing.

Types of Soil

We should realize that the effective removal of complex food soil is a problem to be solved, as are all problems, by finding the right solution. Having accepted this premise, we can then understand that the more we know about the problem, the easier the solution is apt to be. For this reason, we should learn a few basic facts about several distinct types of soil which may be involved in our cleaning operation.

1. Freshly Deposited Soil

   Often the easiest to deal with if removed promptly, since it is the soil remaining after immediate use of utensils and equipment.

2. Thin Film

   A build-up or film is often difficult to detect yet may be capable of sustaining germ life. Dishes coated with film lose their lustre, become stained and require time-consuming soaking plus special destaining and cleaning operations. Glasses no longer sparkle. Pots and pans feel unclean to the touch, accumulate flavors of various foods.

   This film soil may occur for any of several reasons: ineffective cleaning methods, use of the wrong detergent, flushing with plain water, etc.

3. Built-up Deposits

   Again, the employment of ineffective cleaning methods permits a day by day accumulation of soil. Frequently, minerals are incorporated with the various food materials and a hard, difficult to remove soil results.

4. Dried Deposits

   These soil accumulations are the result of drying action and frequently form a heavy, crusty deposit.

5. Baked Deposits

   In operations employing hot air, deposits are baked onto equipment and become difficult to remove.

6. Burned Deposits

   High heat frequently burns material onto equipment surfaces and specific cleaning methods should be employed in its removal. Obviously, the soil to be removed will vary with the type of establishment. Usually, however, a variety of soils will be involved in any given operation. For this reason, several specific chemical detergents should be available to secure over-all satisfactory results.

Water Temperatures

The maintenance of various specific water temperatures during both manual and machine dishwashing is stressed by health agency foodservice ordinances, by sanitarians and by dishmachine and detergent manufacturers and their representatives.

Dish crew personnel who understand why these temperatures are considered proper and necessary will be much more concerned about maintaining them.

One soil encountered in food establishment dishwashing is fat. The various fats must first be softened before the combination of wash solution and friction makes complete removal possible.

Here is a table listing the common fats, together with their melting points.

Type of Fat or Oil	*Melting Point*
Butter fat	95°-107°F.
Cotton seed	41°- 51°F.
Lard (outer fat)	86°-104°F.
Lard (outer leaf)	111°-118°F.
Tallow (beef)	104°-122°F.
Tallow (mutton)	111°-122°F.

We can understand from this table that in washing utensils and equipment by hand, a wash water temperature of 125°F. is necessary for the removal of fats.

Hot water also increases the chemical activity of the detergent being used, thus permitting it to work at the peak efficiency intended by its manufacturer. Handwashing compounds are formulated to act effectively at the wash temperatures specified for this type of cleaning.

Water temperatures given for machine washing and rinsing can be much higher, since the hands are not involved with the wash solution. Different types of detergents, too, are formulated for use in the higher temperature water.

A wash solution consisting of unheated water plus a detergent specially formulated for use at this temperature has proved of value in a variety of specialized cleaning operations. An example is the cold-water glass-washing machine.

Note: In any situation involving the installation of recently developed equipment or the use of a new type product, it is suggested that the local health department be informed. The existing code regulating the sanitation of eating and drinking establishments may or may not cover all the properties of the particular type of equipment or product. Regulations may vary in different localities.

Sanitizing temperatures will be discussed in the section on that subject.

Detergents

When dishes and utensils are correctly washed, sanitized and stored in accordance with health department requirements and the advanced thinking of today's management, we invariably find three factors combining to bring about this happy result: (1) personnel, (2) detergents, and (3) equipment (the tools used for the job, including everything from brushes to dishmachines.)

What are detergents?

For our purpose, we will define detergents as a very large group of cleaning agents, some compounded for manual washing or cleaning operations, others for use in dishmachines, etc.; and these agents, when used correctly, are effective aids in removal of soil from dishes, utensils or equipment.

Leading detergent manufacturers properly regard their products as "aids"—not as miracle workers guaranteed to accomplish desired results in the face of constantly disregarded use instructions, malfunction of equipment or careless personnel.

Specific use instructions are routinely provided on every detergent container label or in a brochure.

Additional details and explanations can always be provided by the detergent company representative.

The first step in good dishwashing procedures is an efficient scrapping set-up.

The final product of good dishwashing procedures is the necessary supply of sparkling glasses, dishes, trays and silver ready for each scheduled meal.

MANUAL DISHWASHING QUIZ

1. Fill in the following statement:
In manual dishwashing and pot and pan washing where washing and sanitizing are done in sinks, a ________________compartment sink set-up should be used. Sink No. 1 for___ _______, No. 2 for ____________, and No. 3 for ____________.

2. Check the correct statements:
(a) Pre-rinsing or scrapping will keep the dishwater cleaner. ( )
(b) A dish rag may be used to wash dishes by hand. ( )
(c) The rinse water in sink No. 2 should be allowed to overflow during washing and sanitizing of both dishes and pots and pans. ( )

3. True or False (check True or False):
(a) Three compartment sinks must be large enough to permit complete immersion, during sanitizing treatment, of tableware, equipment or utensils. Hot and cold running water should be available at all 3 sinks. True ( ) False ( )
(b) A water temperature of 150°F. is used in hand dishwashing. True ( ) False ( )
(c) To measure the detergent accurately, it is necessary to know the amount of water in the wash sink. True ( ) False ( )
(d) After rinsing, pile dishes and flatware in sanitizing solution in sink No. 3. True ( ) False ( )

4. Fill in the following statements:
(a) If a dish is ____________, it cannot be effectively washed or sanitized.
(b) After pots and pans are scrapped, they are often left to ____________ in sink No. 1.
(c) The easiest soil to remove is that which has been ____________ ____________.

5. Check the correct statements:
(a) A thin film build-up is easy to detect. ( )
(b) A temperature of 125°F. will melt fats. ( )
(c) Some dishwashing compounds are made especially for manual dishwashing. ( )

6. True or False (check True or False):
(a) The temperature of the clean rinse water in sink No. 2 should be 100°F. True ( ) False ( )
(b) Glasses of unusual shape or intricate design are easy to wash. True ( ) False ( )
(c) Silverware should be pre-soaked. True ( ) False ( )

7. Check the correct statements:
(a) Pots and pans with film build-up will accumulate flavors of different foods. ( )
(b) Poor cleaning methods will allow soil to accumulate day-by-day until a built-up deposit exists on dishware, pots and pans. ( )
(c) One good detergent can take care of every type washing and cleaning task. ( )

8. Fill in the following statements:
(a) Stained dishes or those with dried-on or baked-on food soil will need ____________.
(b) A clean wash solution has a lower __________________ population build-up.

9. True or False (check True or False):
(a) The backs of dishes must be as clean as the front surfaces. True ( ) False ( )
(b) Improperly washed dishes or utensils can still be effectively sanitized. True ( ) False ( )

10. True or False (check True or False):
Simple methods of feeding and metering liquid detergents, in required amounts, directly from original containers to the wash sink have been developed by leading detergent manufacturers. This modern development has been very successful when installed, used and maintained in strict accordance with the manufacturer's instructions. True ( ) False ( )

MACHINE DISHWASHING DETERGENTS

Notes for Instructors

Here we have a discussion during which we can become too involved in technicalities. Indeed, it was once customary for the instructor to use many chemical terms in that part of the dishwashing session devoted to detergents.

Synthetics, suspension, dispersion, deflocculation, saponification, etc. are all terms often referred to during management-salesmen or health agency talks. No matter how well presented, however, the use of these words during a class attended by employees seems to lead only to confusion, boredom with its resultant inattention and failure to remember clearly, if at all, the necessary basic facts of the discussion.

On the other hand, employees must learn enough about detergents to develop a healthy respect for these very important tools of their trade. They should know something about how detergents work. The simple statement "Use the detergent only as instructed" has little effect in this case. No lasting impression has been made; no significant reason for avoiding careless usage has been given.

If the information is presented in primer form, using only the simplest chemical phrases, past experience in working with foodservice personnel has demonstrated that employees can understand the fundamentals of detergency. The instructor should realize, however, that a certain amount of over-simplification can scarcely be avoided in this type of presentation.

A few unfamiliar terms should be explained to dishwashing employees when their understanding of such terms will lead to a better grasp of the subject. For example, we believe "formula" is a term which should be explained.

Foodservice personnel can often help each other understand how detergents work, even though the instructor has failed to make his point with each class member.

As an example, during an after-class rehash of the material on compounds one member of a dishwashing crew was overheard confessing that "the wetting of water was just crazy talk to him." A fellow worker straightened him out with no difficulty.

"The water starts out being tense," the knowledgeable one explained, "just like everybody gets tense sometimes. So they put some kind of chemical in the water to relax it, just like doctors often need to give tense people medicine to relax them. I know when I'm relaxed I feel better and can work much better. When water gets relaxed, it can do a better job too."

Although "relaxed water" is certainly not a term usually applied to the action of a wetting agent, it still made very good sense to this particular individual.

This conversation has never been forgotten. It would seem to indicate that the instructor can feel free to use any applicable terms, examples or illustrations which make the subject interesting and understandable to personnel.

8 Dishwashing Compounds

1. General

(a) During the hand dishwashing discussion, we referred to detergents as one of the "big three" necessities of a successful dishwashing operation. The other two factors, you will remember, were personnel and equipment. Now it is time to learn something about the work done by the detergent dishwashing compound.

(b) Dishroom crews may wonder why management selects a particular product for establishment use or how it can possibly choose wisely from the many varieties demonstrated by detergent manufacturer representatives.

(c) Management's goal is to find a dishwashing compound best suited to meet the needs of the operation. Taken into consideration are:

(i) The type of establishment and variety of foods served.

(ii) The volume of customers.

(iii) The ability of the compound to best accomplish, under existing conditions, a number of purposes essential to satisfactory cleaning of the dishware.

(iv) The "make-up" of the compound insofar as its action on employee's hands, eyes or clothing is concerned, as well as any undesirable effect it may have on the equipment and utensils to be cleaned.

(v) The high quality of the compound in comparison to its cost. In other words, management searches for the very best in both quality and economy, realizing always that budgets must be maintained.

Item (v) is included here only because dishwashing personnel, by the care with which they carry out dishwashing procedures, often influence management's opinion of a product's economy.

(d) The manufacture of detergents is a complex process. Detergent chemists work out many different combinations of ingredients designed to cope successfully with types of food soil such as those we listed under manual dishwashing. These chemical combinations are tested in the field under actual working conditions, their performances are studied and they are often changed many times until the best results have been achieved. These final results are marketed; management chooses from them.

2. The Dishwashing Compound

(a) Modern dishwashing compounds contain a number of ingredients blended and proportioned to perform their cleaning functions with a minimum of trouble and a maximum of speed, effectiveness and efficiency.

(b) To the majority of dishmachine operators, the abilities of these products are not well understood. Too often they are taken for granted, carelessly used, frequently wasted, yet heartily condemned should they fail to produce the desired results.

(c) When the dishwashing operation does result in spotless, beautifully cleaned dishware, personnel receive management's rightful approval. With no thought of belittling the conscientious efforts of any operator, it is suggested that a glimpse of what is going on inside the dishmachine will give dishwashing personnel a better idea of how much their detergent contributes to the success of their procedures.

3. Important Actions of the Dishwashing Compound

(a) Water Treatment

(i) Since water is the agent combining with our detergent to form the wash solution, we need to know what effect it may have on the performance of the dishwashing compound.

(ii) An accurate analysis of the water supply can be obtained from sources such as a health department or chemical laboratory. In many cases, the detergent representative can provide this information inasmuch as his specific product recommendations will be based on a knowledge of the make-up of the water with which the compound will be used.

(b) Hard Water

(i) Water in various parts of the country possesses many different characteristics but the item most commonly involved with dishwashing is the degree of water hardness.

(ii) When water travels some distance through underground or in surface rivers, lakes, etc., it contacts rock and dissolves into itself various minerals from the rock. Those minerals of chief concern to the dishmachine operator are the lime and magnesium salts, although iron salts may increase staining problems.

(iii) These mineral salts interfere with the cleaning action of the detergent. Certain chemicals in detergents can remove these lime and magnesium salts from the water by throwing them out of their dissolved state into a solid form which does not dissolve. When this happens, the dishwashing operation's efficiency is greatly impaired. There will be a gradual build-up of scale in the interior of the machine, the dishware will be dull and filmed; water spots and streaks will be present on dried dishes, glasses and silver.

(iv) More complex chemicals are able to change the nature of lime and magnesium so that they do *not* combine with other substances, such as the fats and oils so common to food soils, and form a solid substance incapable of dissolving.

(v) Therefore, one of the most important functions of our dishwashing compound will be to keep mineral salts dissolved in the water rather than permitting them to be thrown out of solution to cause scale and filming problems.

(vi) This is the main reason for an accurate analysis of the water supply. Most good dishwashing compounds are available in different grades, to control varying degrees of water hardness. A knowledge of the water's contents is invaluable to the detergent representative in making his product recommendations, and to management in deciding on the most suitable dishwashing compound.

(vii) Although emphasis has been placed on the hardness factor, water may often contain other types of dissolved solids capable of causing dishwashing and rinsing problems. Detergent chemists are already testing various combinations of complex chemicals in an effort to find just the right solution.

Such research paints a bright picture for tomorrow's cleaning routines. Eventually, detergents will be developed to control any problem-causing characteristics of water.

(c) Soft Water

(i) During discussion of hard water problems, some member of the class will invariably ask, "Why don't they give us soft water to work with?" And then somebody else will mention rain water, probably the best known example of a soft water. The answer is that constant use of true soft water would be a mixed blessing.

(ii) It is a fact that true soft water is the exact opposite of hard water in that it contains no dissolved solids. Therefore, no spots or streaks would be left behind on dried dishware as the result of dissolved solid material being changed into a visible substance incapable of dissolving.

(iii) But true soft water is an efficient dissolving agent in itself. Equipment, such as the interior of the machine, and dishware continually exposed to true soft water could age faster than it otherwise would. This means that, in such a situation, the detergent must contain ingredients to control the tendency of true soft water to eat away at and dissolve the materials it contacts.

(d) Softened Water

(i) For dishwashing purposes, softened water is simply water which the dishwashing compound has conditioned (softened) and maintained in this condition without causing dissolved mineral salts to be thrown out of the solution into unwanted solids.

(ii) Water-softening equipment is quite frequently used in an establishment, particularly in areas where the water is of extreme hardness. When water-softening equipment is available to condition water for establishment dishwashing, the detergent selected will have been formulated to work effectively in pre-softened water.

A detergent formula could be compared to a medical prescription. In one case, a doctor lists a number of drug ingredients which, when mixed together in certain proportions and taken by the patient in the amount directed, will help correct a physical problem. A doctor's prescription is taken to a druggist to be filled. The directions for taking the medicine are printed on the bottle or package.

In the other case, a chemist lists a number of cleaning ingredients which, when mixed together in the proper proportions and used in the directed amounts, will be the means of over-coming dishwashing problems. The cleaning chemist's list is called a formula. Its ingredients are mixed correctly by a manufacturer of detergents. Directions for using the detergent are printed on the package.

If the dishroom crew ever expects the detergent to work effectively, it is as important for them to follow the directions on the detergent package as it is to take the proper dose of medicine. Medicines may be in liquid, pill, or powder form. Detergents, too, may be liquids, powders, and, sometimes, solids. This will be discussed later.

Finally, the druggist or pharmacist will have hundreds of different prescriptions on file. The detergent manufacturer also has many formulas available, each intended to solve certain dishwashing problems.

When we say, then, that a detergent has been "formulated" to work best under certain conditions, we mean that a cleaning chemist, knowing what those certain conditions are, has written up a formula containing ingredients capable of controlling the conditions.

4. Wetting

(a) Foodservice personnel give little thought to the varied abilities of water. They all agree, however, that anything (hands, cloths, dishware, etc.) placed in water will become very wet. Knowing this, they will be surprised and interested to learn that water can be made even "wetter" by certain chemicals in the dishwashing compound. In fact, it has to be if the dishes are to be properly cleaned.

(b) Water actually possesses a sort of elastic-like quality which enables it to tighten up or cling to itself. This is why we see drops of water on a dish, drops of dew on a blade of grass, drops of water clinging to the hands, etc.

(c) This same quality prevents water from spreading completely and evenly over and into other surfaces, a condition which must be corrected before the wash solution can work effectively on the removal of food soil. Grease, for instance, resists the entrance of ordinary water which would remove the grease from a plate or utensil.

(d) The dishwashing compound must, therefore, contain special ingredients which will lower what we call the surface tension of water. Then the water and dishwashing compound, or the wash solution, as we call it, can contact every area of the surface to be cleaned; it will enter through the small cracks or openings in the food soil, reach the surface of the dirty dish and push on under the soil to loosen it from the dishware. This action permits all ingredients of the compound to work effectively on the soil, remove it from the dishware and hold it in the wash solution until rinsed away.

(e) Personnel should remember that:

(i) Each ingredient, or chemical included in the compound has been placed there to carry out one or more specific action.

(ii) Not only must the chemicals work together as an efficient cleaning team, they must also get along with each other. No undesirable chemical reactions should occur to interfere with the team's performance.

(f) Now that the water has been softened and made properly wet, we will examine some other important duties of the dishwashing compound's ingredients.

5. Fats, Oils and The Dishwashing Compound

(a) Fats and oils are food soils familiar to every member of a dishwashing crew, including pot and pan washers. Also well known is the fact that these soils have a natural distaste for each other's company, and are unwilling to mix together except on a temporary basis (such as being shaken together vigorously, in which case it takes a little time for them to separate again).

(b) One of the most necessary functions of certain chemicals in the dishwashing compound is to overcome the natural dislike existing between oils and water and permit them to mix together in the wash solution.

(c) The way this is done is interesting. The oil and water do not dissolve in each other. Instead, the oils, or melted fats, are broken up into very fine particles or droplets which then become evenly distributed throughout the wash solution. This is called emulsification. The oil and water remain in this state until flushed off by the rinse water. **Note**: You need not memorize the word "emulsification." It is mentioned because "emulsion" and "emulsify" are quite common words. For example, certain widely used home medications, available at your drugstore, state on their labels that they are emulsions and you may have been curious about the meaning of the word.

(d) Like most cleaning actions, this process is aided by heat. Personnel will recall that, under Hand Dishwashing, some time was spent in a study of the melting points of common fats and oils, with special emphasis on the relationship between these melting points and dishwater temperature.

(e) The higher water temperatures used in machine dishwashing cause solid fats to melt more readily into a liquid form. Thus they can more easily become distributed throughout the wash solution.

(f) The oils and fats thus taken into the wash solution will drain cleanly from the dishware, leaving no oily smears.

6. Solid Soil and the Dishwashing Compound

(a) There may well be instances where solid soils are left on dishware after scrapping and pre-washing, especially if these preliminary routines have been carelessly performed.

(b) The remaining solids must, then, be removed from the dishes and lifted into the cleaning solution. The result is not the same as that accomplished with the liquid oil-water mixture because solid soils may include materials which cannot be completely melted and changed into a liquid form.

(c) A variety of chemical ingredients in the dishwashing compound will break the solid soils up into many very small pieces. Then these tiny pieces will be held up in the wash solution until they can be rinsed away.

(d) The holding up or suspending of particles in the wash solution is another very important function of the dishwashing compound. The reason is easy to understand. If pieces of food soils were allowed to settle, they could form deposits and would not be rinsed completely away. The cleaning operation would not be good because the soil particles would simply be moving from one piece of the dishware to another piece during the wash cycle.

7. The Formation of Soap During the Wash Cycle

(a) Centuries ago ancient man's way of life did not include the high sanitation standards and procedures we practice today. Nevertheless, he shared one problem with modern dishwashing personnel–grease.

(b) Our prehistoric ancestors killed their meat with primitive weapons, cooked it in sizable pieces over campfires and held it in their hands to chew the meat off the bones. Probably their foremost cleaning problem was the removal of the remaining grease from their persons. Slippery hands could not maintain a firm grasp on weapons.

(c) Then, probably by accident, they discovered that when greasy hands were rubbed with a mixture of water and warm wood ashes from their fires, the grease was removed. Ancient man had learned how to make a crude kind of soap.

(d) Today, a situation very much the same exists during the wash cycle. We have:

(i) Grease, as one of the common food soils.

(ii) Hot water.

(iii) Chemicals in the dishwashing compound which act on the grease just as the wood ashes-water mixture did long ago.

(e) This action, causing grease or fats

to emulsify (break up into small particles and remain separated) in water and thus be easily removed, is an aid to cleaning, provided the water hardness has been controlled. Otherwise, this chemical action would result in the soap being changed to a solid which could not dissolve.

(f) The soap formed in the dishmachine's wash tank by the combination of fat, hot water and the alkali chemical content of the detergent will result in a varying amount of suds. A small amount of suds causes no problem; but too much foaming can lessen the wash pump's ability to do its work properly and can act as a cushion between the wash water and the dishware. In machine dishwashing, a good wash jet pressure provides the necessary friction when wash water strikes the dishware with the required force. This, of course, is an important part of the washing operation.

(g) The control of suds can be accomplished by including a chemical in the dishwashing compound which will prevent its formation. Many compounds do contain such a chemical. In food establishments where the machine and the wash solution must handle more than the usual soil load, these detergents are especially valuable. Many good dishwashing compounds are also "non-foamers" in themselves, in addition to possessing the ability to control too much foaming.

The dishwashing crew can play an important part in the controlling of unwanted foam. When the scrapping and/or pre-rinsing is carefully done, most of the grease on tableware will be removed before contact with the wash solution. And if the wash water is changed and renewed often enough, grease should not accumulate in large enough amounts to cause a foaming problem.

(h) We should note here that many of the modern cleaning compounds, so carefully formulated in the laboratory to meet today's cleaning needs, were only a few short years ago the objects of great concern to the cleaning materials industry.

To understand why this occurred, we must think for a minute about wastes.

When the dishwashing crew empties the dishmachine, they know the wash solution disappears down the drain. It probably never occurs to them to wonder where it goes or what, if anything, happens to it.

Thousands of communities take care of the removal of community waste water and refuse by means of sewage systems. Wastes of all kinds are flushed or discharged into sewer pipes, passed along into sewage disposal plants and treated to make them safe and non-objectionable. Then they flow out to mingle with the waters of lakes, streams, or other bodies of water.

There are also communities without sewage treatment systems whose waste water and refuse are dumped "raw," or untreated. In some areas, manufacturing plants have been permitted to discharge their untreated industrial wastes into nearby streams.

Hard water salts as well as bacteria living in the water and soil act continuously on waste matter to change it into different forms. But they are not always successful. This was the case with some of the complex cleaners. Tiny, basic parts of certain chemicals resisted all methods which would have changed them into easily broken down, simple, harmless materials.

When water contains quantities of such wastes, it is said to be polluted, or unclean, or contaminated. Fish and water fowl become sick and probably die. The water is unfit for swimming or any other kind of public usage. Sometimes it becomes discolored.

Detergent wastes became involved in the pollution problem. Wherever large amounts of certain chemical cleaner wastes were regularly discharged into the water, their presence was easily noticed because of foam on the water's surface. The cleaning materials industry was asked to cooperate by changing the make-up of the products involved so they would no longer pollute the water.

Eventually legislation was passed by various states making the necessary product changes into law. Well before this happened, however, leaders of the chemical cleaner industry were moving strongly to do their part towards purifying the public waters.

This involved changing hundreds of formulations so that all parts of the chemicals they included could, when they became waste matter, be broken down into the required non-objectionable forms. These changes had to be made without lessening the cleaning power or any other desirable quality of the products concerned. The amount of laboratory and field test work involved was tremendous. But the end result was that the chemical cleaning industry has made cleaning materials available which will not become polluters of water after their discharge as wastes.

Such products are said to be biodegradable. Dishwashing personnel are not expected to learn this word for routine use. It is mentioned only because they may notice it stamped on certain product labels.

Water conditioning, a very important property of a dishwashing compound, has created great interest among ecologists, researchers and other individuals concerned with the environment—from whatever areas.

In the past, water conditioning was achieved through the use of complex phosphates. When the question of possible phosphate pollution arose, industry chemists began research into different combinations of chemicals which could be developed into products with very few, if any, phosphates.

The use of substitute chemicals has now become widespread throughout the country. But much of the current development in the field of machine warewashing continues to center on the creation of new data and formulations, which will provide 100 percent satisfactory results without endangering the environment.

8. The Dishwashing Compound and Stained Tableware

(a) China, glasses, and silverware are the three types of ware usually thought of in connection with dishmachine washing, although trays may often be run through the machine. Every type of foodservice installation has probably had varying degrees of experience with stained tableware, with the different kinds of "dishes" being, in most cases, the number one offenders.

(b) Several sets of circumstances can contribute to dish staining.

(i) The type of dishware.

(ii) The dishwashing procedures.

(iii) The type and concentration of the dishwashing compound.

(c) Under "types of dishware" are included:

(i) China or porcelain, the most common or standard dishware of past years and probably still the most widely used;

(ii) A glass tableware, resembling chinaware but possessing different physical qualities (example—milk glass);

(iii) Plasticware, remarkably improved and, hence, increasingly popular since its introduction, now meeting the needs of many foodservice establishments.

(d) (i) In the manufacture of china, the final step is to cover each piece of ware with a shiny, glass-like material called a glaze. When this glaze is baked on, it becomes a very hard and protective outer coating for the dishes.

(ii) If the shiny glaze becomes dull and stained, supervisors should check both the dishwashing routines and the compound. No step in the washing operation should be neglected or hurried through by personnel. The detergent dispenser should be checked to be sure it is working properly, with a sufficient amount of the dishwashing compound maintained in it at all times.

(iii) As for the compound itself, there is always the possibility the product may not be exactly right for the particular operating situation. Every cleaning materials manufacturer offers a variety of dishwashing compounds; and here, the detergent field representative can be of great assistance. Another of his products, with, perhaps, only, a slightly different chemical make-up, may be the answer.

If a rinse additive, which we will discuss shortly, is being used, it, too, should be checked to be sure it is the proper one for use under operating circumstances.

(iv) A problem with the dishwashing procedures and/or the dishwashing compound could very likely result in the build-up of a thin film on the dishes.

Film build-up was covered in general terms under "Types of soil." Now we think of it in connection with dishstaining.

(v) Film build-up on dishes will usually be gradual, increasing with each trip of the ware through the machine. In the beginning, its presence may not be noticed. It does not form a hard shell on the dishes as, for instance, the china glaze does.

Rather, *this* film is full of tiny openings or pores like those of the skin. Food soils can be soaked up in these pores and remain in them as unsightly stains on the dishes. When this happens, dishwashing crews are usually instructed to soak the dishware in a bleaching solution. This practice may correct the condition temporarily, but will not remove the cause. Whatever is amiss in the operation will lead to still another build-up of film.

(vi) There is yet another reason for the appearance of stains on china. The lifetime of this ware is long. But daily washing, handling and usage finally will cause the china's hard, protective outer glaze to lose its smoothness; damage from knife scratches, etc. cannot be avoided; careless handling will result in cracks and chipping. Thus, with its outer glaze marred, cracked, or worn off, food soils can gain entry into the inner surface of the dish and become unappetizing appearing stains.

(vii) Effective washing becomes impossible under these conditions. Growing places for bacteria are provided in the worn and/or cracked or chipped areas of dishes, thus creating a problem in effective sanitizing.

(viii) Neither an expensive dishmachine nor use of the proper cleaning materials can guarantee everlasting life for tableware. China and other types of ware cannot avoid being affected by long, continual usage in busy foodservice establishments. For both public health and esthetic reasons, the good policy of modern management dictates the immediate replacement of damaged or worn-out items of tableware.

(e) Plasticware

(i) When plastic dishes were first placed on the market some years ago, the ease with which stains accumulated on the ware was an annoying problem. Since then, however, plastics have been tremendously improved. And while the need for control of stain still exists, as it does with other types of ware, it is no longer the urgent problem of those first days of the plastic dish.

(ii) The cleaning materials industry developed dishwashing compounds of a different kind to help control staining on plastic, and other types of tableware. These were called bleaching or chlorinated dishwashing detergents. Dishwashing crews in establishments where plasticware is used will be familiar with these detergents, as they are most often recommended for stain control. They may, however, be found in other areas not concerned with plastics. The washing of glassware, is one example.

(iii) Chlorine, which is contained in these detergents, is a chemical, well known as a bleach. When used in the required concentration (solution strength), it is also the most widely used of the health department-approved chemical sanitizing agents. We will learn more about chlorine when we come to a discussion of sanitizing.

(iv) In some instances, management and personnel believe the presence of chlorine in the compound lessens the build-up of film on dishware. Knowing of chlorine's value as a sanitizer, they also like to use these compounds routinely for any possible benefit the chlorine could have in keeping the machine's interior in a fresher condition, as well as in lowering the bacterial population of the wash solution. Hence, the chlorinated dishwashing compound may be selected by management for a wide variety of uses unrelated to plasticware.

(f) It is common practice to discard cracked or chipped glasses. The surface of glassware, however, may become scratched. While not exactly a staining problem, it should be noted that, once scratches have occurred, continuous washing of the glass in a hot water-detergent solution can speed up the development of an over-all web or design of scratchy marks which makes the glassware no longer presentable for use.

Manufacturers of glassware, constantly improving the make-up and design of their products, are very conscious of modern management's requirements that glassware be beautiful, but also cleanable according to sanitation regulations. The unusually shaped, practically impossible to clean glass of former years is a rarity today. The glass designed so that every area of its surface may be effectively contacted by wash solution and rinse water has become a standard item.

(g) Tarnish or stain on flatware may

be removed by several methods, depending on the metals of which the utensils are made. Silverplated knives, forks, spoons, etc. are widely used. German silver, actually a combination of copper and nickel, containing no silver is available; and stainless steel utensils have become popular in many installations.

Information and instruction on the proper detarnishing method should be made available by the manufacturer or flatware salesman, although the detergent representative can undoubtedly offer valuable advice. Most dishwashing personnel are familiar with the mechanical burnisher; have used a destaining compound in a sink of hot water with (depending on the chemical make-up of the product used) a crumpled piece of aluminum foil or clean, flattened No. 10 tins, or similar items placed in the bottom of the sink to contact the flatware during the necessary soaking period.

After long usage, stainless steel ware may show a bluish tinge which can usually be removed in a mildly acidic dip. A variety of special destaining products is available from chemical cleaner representatives.

(h) Aluminum trays

(i) When their size and shape permit, aluminum trays are often run through the dishmachine. A problem may be created here because many of the finest dishwashing compounds, usually problem savers in themselves, are not suitable for washing aluminum. Their use can mar the tray's appearance by a slow chemical action of "eating away" at the tray's surface.

(ii) Rough spots may then develop on the trays; a kind of splotchy white rust may appear; or black smudges may be produced which will wipe off on uniforms, towels, or the hands of personnel. If the trays are used in cafeteria service, the hands and clothing of customers may be soiled.

(iii) This situation may be corrected by washing the trays separately by hand, using a proper detergent. Or the dishwashing compound may be changed to a completely inhibited one in which the undesirable chemical reactions are totally prevented from occurring, thus making it safe for use on metal.

(iv) Circumstances, like an unusually heavy soil load, or preference for a chlorinated compound, may make management decide to continue the use of the current dishwashing compound, even though it is not formulated to prevent an undesirable attack on aluminum. If the attack is not too severe, the trays washed in the machine will very likely require more frequent replacement than normally, but may still be usable for some time.

9. Rinsing

The wash solution must flush easily and completely from the surfaces of the dishware. This is accomplished when all of the chemical actions we have been discussing are working together properly:

(a) Softening the water

(b) Reducing its surface tension (wetting)

(c) Keeping the food soils well suspended in the wash solution.

10. General

(a) Our dishwashing compound must be able to dissolve easily in the concentrations (strengths) needed to thoroughly cleanse the dishware.

(b) The compound should have no harmful effects on the surfaces of the dishes or utensils being cleaned.

(c) It should not be irritating to the skin if hand contact with the wash solution is necessary.

(d) The compound must satisfy the requirements of ecology. It must not pollute vegetation, water, etc., or endanger animal life, fish or other living components of the environment.

By now, dishwashing personnel should have a better understanding of the many important activities being carried on by the dishwashing compound's ingredients. They will know their compound contains chemicals which will perform all these duties well, according to the kind of soil to be removed from the dishware. They will realize that no single chemical ingredient can perform all of the necessary operations successfully; therefore, it is necessary to combine the proper ingredients so that each cleaning activity may be carried out most effectively.

DISHWASHING COMPOUND QUIZ

1. Fill-in.
The "big three" of a successful dishwashing operation are ________, ____________, and ______________.

2. Check correct statements:
(a) A modern dishwashing compound should work with speed,effectiveness,and efficiency. ( )
(b) Hardness of water is no problem in dishwashing. ( )
(c) The continued use of true soft water could cause trouble with dishware and equipment. ( )
(d) A well formulated dishwashing compound will soften water in addition to its other functions. ( )

3. True-False (check True or False):
(a) The dishwashing compound is made up of different chemicals, which work together to thoroughly clean the dishware. True ( ) False ( )

4. True-False (check True or False):
It is possible for water to be wetter. True ( ) False ( )

5. True-False (check True or False):
Hand dishwashing compounds and machine dishwashing compounds are usually formulated with the same chemicals. True ( ) False ( )

6. True-False (check True or False):
Solid fats melt more readily at higher temperatures. True ( ) False ( )

7. Fill-in
(a) Solid soils must be broken up into tiny __________, and kept ___________ in the wash solution so they will not settle on the ______________.
(b) The first soap was probably a simple mixture of ____________, plus ____________ from a camp fire, and _____________.

8. Check correct statements.
(a) Knowing what is expected of a dishwashing compound will help the dish crew use it more carefully and efficiently. ( )
(b) Accurate measurements of detergents is not actually necessary. ( )
(c) An analysis of the water supply helps in the selection of the most suitable dishwashing compound. ( )
(d) Use of a good dishwashing compound makes thorough scrapping unnecessary. ( )

9. True-False (check True or False):
Each chemical in the detergent has been included to perform one or more specific actions. True ( ) False ( )

10. True-False (check True or False):
The same dishwashing compound can work equally well in any foodservice operation, no matter where in the country it is located. True ( ) False ( )

9 Detergent Dispensing Systems

Before discussing how various dishwashing compounds may be dispensed, we will review briefly, the different forms in which these detergents are available.

(a) Powder (granular) form:

Detergent compounds in powder form are the "old-timers" of the foodservice industry. Over the years, they have developed from simple chemical formulas containing only the basic cleaning ingredients to the complex products of today, performing a variety of tasks. Many of these were previously covered in our detergent discussion. It is probable that the powdered compound is still the most widely used in the foodservice field.

(b) Liquid form.

Liquid compounds are becoming increasingly popular. They, too, have been greatly improved and can perform very efficiently in the many situations where a liquid detergent is considered to best fulfill the particular purpose for its use.

The liquid detergent is not a mere solution of powdered detergent materials, though many personnel believe it is. It has, rather, been specially manufactured in its liquid form for the particular types of cleansing where it is regarded as most efficient.

Dispensers for Dishmachines

WHAT THEY ARE–A detergent dispenser is a device for adding detergents to wash water so that the desired strength of the detergent solution is maintained throughout the dishmachine's operating periods.

Dishwashing personnel who have previously discussed the "Dishwashing Compounds" material will find the advantages of a dispenser easy to understand. They already know that:

(a) Their dishwashing compound has been "built" or manufactured according to a particular chemical formula.

(b) A stated amount of this compound must be present in a certain amount of wash water in order to create the proper strength wash solution for effective cleaning of dishes and utensils.

(c) Too little detergent in the wash solution will result in unsatisfactory cleaning. On the other hand, too much detergent in the wash solution is both wasteful and costly. Waste and increased cost due to personnel's carelessness or ignorance are conditions not tolerated by alert, modern management.

For these reasons, members of the crew are not apt to question the "why" of a dispenser's use. They are more interested in learning something about how these devices work, particularly those employees whose dishmachine is equipped with one of the more sophisticated systems.

Background and Development of Detergent Dispensers

The first dispenser was undoubtedly the human hand and this, admittedly, was a most unsatisfactory device. About the year 1935, the custom was to add dishwashing powder to the wash solution when needed.

The problem here was the vagueness of the term "when needed." A cupful or two might be used in the beginning, with another cupful supposedly to be added an hour later. Or the compound might be added every certain number of racks of dishes put through the machine. During rush hours, however, when too busy to refer to the clock or count racks, many dishwashing personnel added their detergent to the wash tank when, as and if somebody in the crew happened to think of it.

If a cup was conveniently handy, it would be used. But often a handful was scooped from the container; or a guessed-at amount was poured from a box. As a result, many dishmachines might operate for some part of the busiest periods with little or no detergent in the wash tank. As might be expected, the results were unsatisfactory.

To correct these conditions, several of the leading detergent manufacturers developed the first hydraulic dispensers, hoping by this means to keep the concentration of the dishwashing compound at a more even level.

The Hydraulic Dispenser

The term "hydraulic" means having something to do with water in motion. This type of dispenser permits additional amounts of dishwashing compound to be fed into the wash tank each time the machine's hot water rinse system is operated.

An open stainless steel container, also termed a hopper or reservoir, is mounted on the machine. It is the responsibility of the dishwashing personnel to see that the reservoir is kept supplied with a sufficient amount of dishwashing compound and that it is clean and in proper operating condition.

Tubing, connected to the pressure side of the final rinse line, leads to the detergent reservoir. Every time the rinse valve is opened, hot water flows through this tube into the dispenser reservoir, dissolves a certain amount of the detergent, and an additional amount of fresh detergent solution is then fed through a second tube directly into the wash tank.

In this way, more dishwashing compound is added at regular intervals to the wash solution without any dependence on the hit-or-miss methods of busy or forgetful personnel.

These hydraulic devices represented a great advance in detergent dispensing. Their development was based on the reasoning that:

(a) The wash solution would be diluted somewhat by each release of hot rinse water.

(b) Each use of the rinse meant that one more load of dishes had been washed; therefore the wash solution would be somewhat weakened by contacting and taking into itself the food soil present on that load of dishes. It was thought that by adding a fresh charge of dishwashing compound to the wash solution with each use of the rinse, these conditions would be corrected.

Before very long, however, it was realized that the hydraulic dispenser did not provide all the answers. The main problems were:

(a) Dishmachine personnel often neglected to keep the detergent hopper supplied with sufficient amounts of the compound.

(b) The addition of small amounts of compound with each use of the final rinse did not guarantee the same, measured detergent strength solution in the wash tank throughout the machine's entire operating period.

(c) Multiple tank, conveyor-type machines were finding their place in the larger installations.

Again, the leading detergent manufacturers met this challenge by developing the first electronic dispensers.

Today some form of electronic controlling device to maintain detergent concentration will be found in wide use in commercial dishrooms. But the hydraulic type also has its place. For example, the National Sanitation Foundation's Standard No. 3 points out that in situations where the work load is fairly light, the hydraulic dispenser can be a satisfactory dispensing tool, especially when an indicating meter is also installed on the machine. This latter device will be discussed under the heading, Detergent Concentration Meters.

The Electronic (Electric) Dispenser

The early electronic dispenser consisted, in most cases, of a detergent reservoir situated on the machine. Jets of hot water, spraying periodically into this reservoir or hopper, would result in an amount of concentrated detergent being dissolved and released into the wash tank.

Many dispensers of this type are still in operation, so a rather detailed explanation follows as to how they work.

1. WHAT IT DOES

This type of dispenser measures the strength of the detergent solution in the wash tank. Whenever the concentration drops below the required level, detergent is added to the wash tank in the needed amount. When the proper concentration has been reached, the feeding of detergent is stopped and the dispenser resumes its measurement of the wash solution's strength until time to add more dishwashing compound to the wash tank.

2. CONSTRUCTION

These devices work by means of electricity. They have three main parts. These are:

(a) A "brain," termed a control head, an electronic solution controller or other equally descriptive name given it by its manufacturer.

(b) An electrode (resembling a prong), installed in the wash tank and connected to the "brain."

(c) A solenoid valve, also connected to the control head. This valve, installed on the pressure side of a hot water line, is also connected to the container which holds the dishwashing compound.

A solenoid is a special type of valve which acts like a magnet when an electric current passes through it. Because it has this ability, it works automatically to open or close whenever it receives the appropriate signal. In other words, it works by itself; personnel have nothing to do with its operation.

3. HOW AN ELECTRONIC DISPENSER WORKS

(a) The detergent representative adjusts the control unit or "brain" to be aware of the desired concentration of detergent in the wash tank. Now, when the dishmachine is operating, the unit will continually be measuring the strength of the wash solution and will be working to keep the detergent strength at the correct level.

(b) The washwater solution containing the detergent is an electrolyte, which simply means it is a solution having the power to conduct an electric current. (In this case, the current is very weak and is not felt by humans.) The pre-adjusted, stated detergent concentration will possess the ability to conduct a certain strength of electricity.

It is easy to understand that when the strength of the detergent solution changes for any reason, there will be a corresponding change in the degree or amount of electrical current which is passing through the wash solution.

(c) To detect these electrical changes, an electrode is mounted through the wall of the wash tank well below the level of the wash solution, Personnel may have heard this electrode referred to as a probe or sensing tool.

This particular electrode has two poles (wires or prongs) which form an electrical connection. Since the electrode is completely covered by the wash solution, it carries an electric current of the same degree as the small flow of electricity passing across the dispenser circuit.

(d) The electrode is continuously measuring the strength of the wash solution and sending this information back to the control head or "brain." When the concentration drops below the pre-set required level, the strength of the flow of electric current will show a corresponding drop. When the control head receives the signal that this has occurred, it sends a flow of electrical current to the solenoid valve, giving it the power to open.

(e) When the solenoid valve opens, a flow of hot water is released into the dish compound reservoir. The hot water and the detergent it dissolves form a concentrated solution which overflows through a tube into the wash tank. Thus, the strength of the solution in the wash tank begins to build up until it reaches the pre-set concentration level to which the control head has been adjusted.

(f) When the concentration reaches the proper level, the amount of electrical current in the wash solution regains its original strength. The electrode passes this information on to the control head. The flow of electricity to the solenoid valve is discontinued and the valve closes. When the valve closes hot water is no longer free to flow into the detergent reservoir and the flow of dissolved detergent into the wash tank stops. This entire cycle is repeated each time the concentration level falls below the pre-set adjustment and occurs many times during the dishwashing operation.

Since the advent of the model described above, advances in technology have resulted in the development of models in which some instrument or agent other than the solenoid valve puts the dispensing process into action.

These agents will be discussed in the section following Detergent Concentration Meters, (p. 77).

(g) A leading manufacturer of detergents and detergent dispensers applied the following terms to the three main parts of the electronic dispenser:

i. The electronic solution controller is the Brain.

ii. The solenoid valve or some other agent activating the detergent delivery, is the Muscle.

iii. The electrode is the Nerve.

This comparison could be carried further by stating that the Brain (control

Electronic Dispenser Control Unit, above, is mounted on top of machine. Receptacle to hold dishwashing compound is mounted on side of machine. Note soiled dish scrapping table with recirculating water in trough and disposer unit.

Electronic Dispenser Control Unit and Rinse Additive Injector, right above, is mounted on top of machine. Note also the scrapping table, where scrapping is done into troughs directly in front of operators and flushed into square sink on corner. This sink is fitted with a waste disposer unit. A flush cover is placed over the sink during operation.

Electronic Dispenser Control Unit, right, is mounted on top of machine. On right, side mounted receptacle which holds dishwashing compound. On left, side mounting of receptacle which holds rinse additive. Piping for both devices carried down and beneath machine, rather than on top. Lower container is for food soil discharged from pre-wash unit.

head) does the thinking; the Muscle (solenoid valve) performs a physical action in opening and closing; the Nerve (electrode) reacts to the changes and in so doing sends electrical impulses to the Brain.

Personnel should find the use of these terms helpful in understanding how the electronic dispenser carries on its work.

(h) One of the problems mentioned in connection with the hydraulic dispenser was the failure of some dishwashing crews to keep the detergent container or reservoir regularly supplied with a sufficient amount of detergent.

This problem has been solved by the installation of some type of warning system on most electronic dispenser reservoirs, its purpose being to let the dishwashing crew know when more compound is needed. A red light, a persistent sound or, in some cases, both, will continue to signal until the situation has been corrected.

(i) Since this dispensing equipment is supplied by the detergent manufacturer, it will be expected that differences will exist in various models. The manufacturer provides the dispenser which will work best with his products. These differences can be explained by the detergent representative. Our discussion is intended to provide dishwashing crews with as simple an explanation as possible of the basic method of operation of this type of dispenser.

(j) Dishwashing personnel employed in large industrial or institutional operations will be familiar with the heavy-duty electronic dispenser models, which are also very effective on pot and pan washing machines.

Detergent Concentrations

When dishwashing personnel think of the detergent concentration as the strength of a solution made up of the dishwashing compound and water, in certain stated amounts of each, the term is easy to understand. Often, however, they hear mention of "desired," "required," "proper," "needed," concentrations and may wonder just what is meant by this. How much detergent to how much water? And how do we know this amount is right for the job to be done?

As long ago as 1944, a concentration of 0.30 had been accepted as, in general, being the most satisfactory. This percentage might vary slightly with different detergents.

The 0.30 percent concentration was accepted after a Public Health Service officer had made a thorough study of washing and sanitizing methods, tested various detergent solution strengths, etc. and published the results of his investigation. His 1944 paper was so complete, so well organized and clearly presented, it has continued to be considered a reliable guide and reference work up to the present time.

An 0.30 or .3 percent solution strength means there should be 3 oz. of detergent for each cubic foot of water in the wash tank. (This proportion rate is also provided by a leading detergent manufacturer.) In terms of gallons, this would be 3 oz. of detergent to approximately 7½ gal. of water in the wash tank.

To keep this solution strength continuously at the same level, more detergent must be added to the wash water whenever, for any reason, there is less than 3 oz. of detergent for every 7½ gal. of water in the wash tank. This is why the various types of detergent dispensers and detergent concentration meters or indicators are recommended for use in modern dishwashing operations.

Since manufacturers of detergents test products thoroughly before they are marketed, the strength solution at which each dishwashing compound operates most effectively is known and recommended for usage. These recommendations should be carefully followed.

The chemical formulations of dishwashing compounds have been greatly improved during recent years, and many specialized compounds have been developed. For these reasons, it is not unusual to find quite wide variations in recommended concentrations. There may be a range of from .15 to .30 percent, depending on the type of compound and the conditions under which it will be used. The detergent representative will know the best concentration for the particular situation and will adjust the dispenser accordingly.

Note: The recommended strength of the detergent solution should be posted in a conspicuous spot on or near the dishmachine where personnel can see it. This is especially important where the compound must be added by hand.

Whenever an employee refills or adds more compound to the dispenser hopper, it is a good idea to have him write his initials and the hour on a sheet of paper posted conveniently near the dispenser.

When dispensers do not operate properly, personnel should never attempt to make any adjustment. The detergent representative should be called at once.

Detergent Concentration Meters

WHAT THEY ARE–A detergent concentration meter is a device which shows the strength of the detergent solution in the wash tank. It may be used with or without a detergent dispenser.

HOW THE METERS WORK–The detergent representative sets the detergent concentration meter correctly for the particular needs of the dishwashing operation. He will also set the electronic dispenser.

Also, as with the electronic dispenser, an electrode allows the meter to indicate continuously the strength, or concentration, of the detergent solution in the wash tank.

An indicating dial is located on the side of the meter to show the strength of the wash solution. On the dial's face a curved band, divided into sections, is plainly marked with explanatory words such as "add compound," "OK," "Overcharged."

Here again, detergent manufacturers who provide these meters use their own descriptive terms. "Hi," "OK," "LO," or "Add," "OK," "Too Much," are other examples of wording which personnel can easily note and understand.

While the machine is operating, the needle on the dial will point to the area indicating strength of detergent solution.

There are several valuable uses for detergent concentration meters. Where detergents are fed by hand into the wash tank, dishwashing employees are guided as to when and how much compound to add by noting to which section of the indicating band the needle is pointing.

The needle should remain in the OK range all during the washing period. If it drops into the Add section, more detergent should be immediately added to bring it back to the OK area. If too much detergent is added, the needle will swing to the "too much," "hi," or "overcharge" area.

Where a hydraulic dispenser is used, the meter assists the operator in keeping the detergent solution at the proper level. He can tell when the dispenser is too low on detergent without having to check it.

Managerial personnel can quickly check for proper detergent concentrations by noting the meter reading.

The concentration meter can also be used as part of a dispensing system from the original shipping container to the wash tank. This dispensing method will be discussed in the following section.

Note: Any type of control equipment needs to be inspected and adjusted periodically by the detergent representative. Where no control equipment is used, management should be provided with a chemical field testing kit for checking and controlling the concentration. These tests are easy to make and can supply the necessary control information for the specific time at which they are made.

Variations in Electronic Dispensing Methods

AUGER SYSTEM DRY POWDER FEEDER

This method also involves a detergent hopper situated on the dishmachine, and containing a powder detergent.

When the dishwashing compound in the wash solution falls below the correct level, the electrode in the wash tank signals the solution controller (Brain or Sensor). The controller sends a flow of electric current to an auger (Muscle) located on top of the machine. The auger forces a predetermined amount of dry, concentrated detergent directly into the wash tank.

DISPENSING DETERGENTS FROM ORIGINAL CONTAINER–This dispensing method provides a convenient, tidy and accurate way of keeping the detergent strength at the proper level. Personnel need not handle the dishwashing compound because the detergent reservoir, or hopper, is eliminated. Thus, time is saved, no detergent is wasted, and there is more efficiency in the dishwashing operation.

DISPENSING FROM ORIGINAL CONTAINER USING ELECTRONIC SOLENOID OPERATED LIQUID SIPHON SYSTEM–Whenever the electrode in the wash tank signals the electronic controller that the detergent concentration in the wash solution is low, the controller supplies power to the solenoid valve, causing it to open. This allows hot water to go through siphon.

The liquid concentrate contained in the shipper (container) is siphoned (drawn) out, mixed with the water and delivered to the wash tank, thus building the solution up to required level. When the solution strength reaches this proper level, the solenoid valve is signalled to close and will remain closed until the solution strength again falls below the predetermined level.

DISPENSING POWDERED DETERGENT FROM THE SHIPPING CONTAINER–There are two ways in which a shipping container may be used in the dispensing of powdered detergents: (1) directly from container to the dishmachine's wash tank; or (2) in a remote control operation, where the container may be situated some distance from the dishmachine in another room, or even on another floor.

In (1), when the electrode signals the electronic controller that the concentration of detergent in the wash solution is low, power is supplied to the solenoid valve which allows it to open. Hot water is admitted into the detergent shipper. The concentrated solution of hot water and detergent is fed into the wash tank, thus building up the solution strength. When the strength of the solution reaches the pre-set level, the solenoid valve closes.

In (2), as above, when the strength of the detergent solution falls below the preset level, the electronic controller energizes a relay which pumps concentrated detergent solution from a sump reservoir to the machine.

In this instance, the shipping container is placed in an inverted position upon a dispensing device. Brief spurts of water jet up into the container, producing a mixture of concentrated detergent and water which is available for pumping to the dishmachine's wash tank.

DISPENSING LIQUID DETERGENTS FROM A REMOTE AREA TO THE WASH TANK–In this case, the electrode in the wash tank signals an electronic controller situated on a wall above the container of liquid detergent.

When the electronic control receives the signal that the wash tank solution has fallen below the required detergent concentration, it activates a piston pump which delivers the detergent from the shipper to the dishmachine wash tank.

DISPENSING DETERGENT FROM SHIPPING CONTAINER, USING DETERGENT CONCENTRATION METER–The detergent representative sets the concentration meter correctly for the particular needs of the dishwashing operation. The electrode connected to the meter keeps the meter informed of the wash solution's strength.

This information is shown on the meter's dial by its indicating needle, which will point to the appropriately marked section on the face of the dial.

Whenever the needle falls below the OK section into the area indicating that more compound is needed, a crew member will open a conveniently located spring-loaded valve on a hot water supply line. The water flows through the shipping container and a correctly proportioned amount of compound is fed through a tube into the wash tank. When the indicating needle again points to the OK section of the dial, the valve is closed.

Personnel should keep a close check on the meter reading in order to know when to open the valve and thus begin the building up of the concentration to its required level.

DISPENSING FROM CONTAINER TO WASH SINK IN MANUAL DISHWASHING AND POT AND PAN WASHING–This is another development which is becoming more and more widely used. Many personnel who wash dishes or pots and pans by hand are familiar with the method. Concentrated liquid detergent is dispensed.

While there are mechanical differences in these various control devices, they operate, in general, as follows: A special type fitting, attached to the faucet used in filling the sink, is also connected by tubing to the drum or container of liquid detergent sitting on the floor.

When the water is turned on and flows through the fitting, the liquid detergent is drawn up through the tube, mixes with the water to form the desired wash solution, and flows on into the sink, ready to use.

A button on the fixture may be pushed in to allow the detergent to begin its flow into the running water. When the water is turned off, the button pops out automatically, cutting off the flow of detergent. With this arrangement, clear water is available whenever the button is not pushed in by the dish or pot washer.

The amount of liquid detergent used may be controlled or metered by various methods, but the end result is the correct proportioning of detergent to each gallon of water to form the required concentrations. Here again, the detergent manufacturer's recommendations should be followed and his representative notifed if repairs or adjustments are thought necessary.

Rinse Injectors

WHAT THEY ARE—A rinse injector is a mechanical device which automatically injects, or forces, a small amount of a chemical called a drying agent into the dishmachine's final rinse just before the rinse water sprays from the rinse nozzles over the tableware. This can be done in various ways, such as by using a pump, suction, or some other means.

WHAT IS ACCOMPLISHED BY USE OF A RINSE INJECTOR WITH CHEMICAL DRYING AGENTS—

1. The rinse water runs off the tableware evenly, in sheets, leaving no spots or streaks.
2. The dishes and utensils dry much more rapidly.
3. There is no need to towel the tableware.

HOW THESE RESULTS ARE ACCOMPLISHED—

1. The drying agents.

These liquid chemicals, also referred to as rinse additives, are automatically fed, in very small amounts, into the final 180°F. rinse water. They lower the surface tension of the water so that it "sheets off" the tableware evenly, rather than collecting in drops which, in drying, leave unsightly spots and streaks on the dishes.

Controlled electronic detergent dispenser maintains proper concentration using container as detergent reservoir. Audio and visual signals indicate any need for additional supply.

This is more easily understood if personnel will remember that it is the wetting agents, used in the dishwashing compound to lower the surface tension of the water that make it possible for the other chemicals in the compound to penetrate through and under food soil to loosen and remove it from the dishes.

In the case of drying agents, the purpose of lowering the surface tension of the water is entirely different. The surface tension is lowered to permit the rinse water to spread evenly and thinly over the tableware. The ware has already been heated during its passage through the machine.

When the 180°F. water spreads over the dishes in a thin sheet, almost like a thin film, the additional heat causes almost immediate drying of the ware. This is why the term "drying agent" is correct when the chemical is used for this purpose.

2. A chemical drying agent—usually termed a rinse additive—is not intended to act as a dishwashing compound and will not remove food soil from dishware.

For this final treatment to be effective, all previous steps in the dishwashing routine must be carefully performed. This includes pre-scrapping, correct wash and rinse temperatures, use of the most suitable compound, proper racking of dishware, an adequately ventilated dish room; plus the selection of the right rinse additive for the operation.

In addition, the rinse injector equipment must be kept clean, properly maintained, and supplied with the drying agent, just as the detergent dispenser has to be kept supplied with a sufficient amount of compound. Again, dishwashing personnel should not attempt to adjust the rinse injector unit.

3. Some drying agents are formulated to work best in hard water areas, while others are most effective in soft, softened, or medium-hard water.

Where deposits of lime scale in the rinse sprays is a problem, the rinse additive formulated for hard water use can be effective in keeping the sprays free of such deposits. The treated rinse water also acts to dissolve any food stains or signs of mineral deposits on the dishes each time they go through the washing and rinsing process.

4. The development of rinse injectors for use with chemical drying agents is considered to be one of the greatest advances made in the field of mechanical dishwashing in recent years. When correctly used, it insures spot-free dishes, elimination of insanitary hand towelling, more rapid drying of tableware with a quicker return to service. All of these are advantages that save time and money for management and help ensure a sanitary dishwashing operation.

Temperature & Food Sanitation

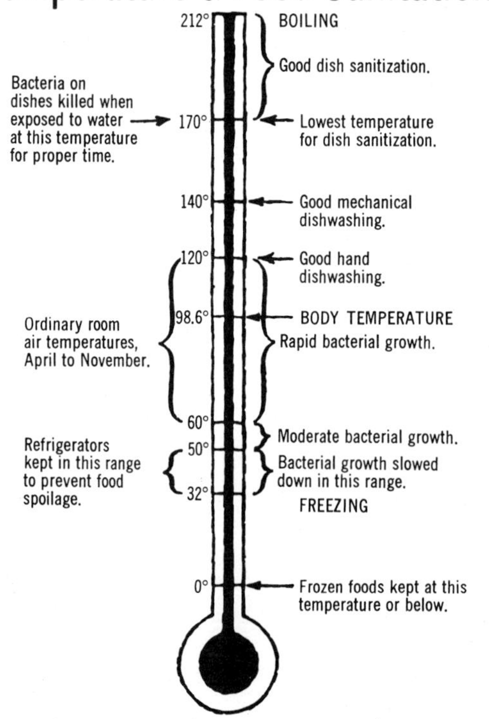

DETERGENT DISPENSER SYSTEMS QUIZ

1. Fill-In
   A detergent dispenser is a device which adds __________ to the ________________.

2. Fill-In
   This helps maintain the _______________ of the ______________ at required level throughout the operating period.

3. Check correct statements.
   (a) The hydraulic dispenser feeds dishwashing compound into the wash tank each time the hot water rinse is operated. ( )
   (b) Hydraulic dispensers are best for large machines and heavy work loads. ( )
   (c) The electronic dispenser works by means of electricity. ( )
   (d) Dishwashing personnel need to adjust the electronic dispenser almost every day. ( )

4. True-False (check True or False):
   A correct detergent concentration is to have 3 oz. of dishwashing compound for every 7½ gal. of water in the wash tank. True ( ) False ( )

5. True-False (check True or False):
   A detergent concentration meter will take the place of a dispenser. True ( ) False ( )

6. Name the three essential parts of the early electronic dispenser.
   1 _______________ 2 _______________ 3 _______________.

7. Name two modern detergent dispensing systems. _______________, ______________.

8. True-False (check True or False):
   A rinse injector is a device whereby a small amount of chemical drying agent is mixed with the dishmachine's final rinse. True ( ) False ( )

9. True-False (check True or False):
   A rinse additive (drying agent) removes soil from dishware. True ( ) False ( )

10. True-False (check True or False):
    A rinse additive aids in fast drying, elimination of spots and streaks, need for toweling.
    True ( ) False ( )

Notes for Instructors

To achieve maximum results from the dishmachine, the dishwashing crew should have a working knowledge of the equipment they are using. The following outlines of dishmachine types are intended to help dishwashing employees become more familiar with various common models and to give non-dishwashing personnel some idea of the scope of the subject.

Employees should be informed that not every type of machine can be covered in such a short discussion and that they may encounter slight variations in machines of the same general classification. Obviously, competitive manufacturers will not produce identical models.

Since the purpose of the instruction is to inform members of the average dishroom crew, as well as to interest other establishment employees, technicalities have been avoided as much as possible. No attempt is made here to explain such items as required pump capacities, degrees of water volume and pressure, plumbing installations or the details of water heating systems. For this purpose, the dishmachine company's representative would be most helpful.

The data plate that, in most instances, is permanently affixed in a conspicuous location on the machine provides valuable information and should be frequently referred to by the dishroom crew. The machine's required wash and rinse temperatures, wash and rinse time periods, the chemical type and concentration of sanitizer (when applicable) and optimum final rinse pressure are all facts necessary to the *proper performance* of daily routines.

All other factors being correct, if the dishes come out of the machine still partially soiled or improperly rinsed; or if the machine's thermometers indicate incorrect temperatures, a dishroom crew member should immediately inform a supervisor who should have the proper individual called at once to correct such malfunctions.

More advanced training on their own dish machine's construction and working system can, of course, be provided at the discretion of management.

10 Mechanical Dishwashing

BASICS

The basic requirements of wash, rinse and sanitize are common to both hand and machine dishwashing. But because a wide range of dishmachine models have been made available to meet the needs of all types of foodservice installations, there are variations in the steps by which these basic procedures are accomplished.

One of the most widely used types of dishmachine in the commercial field is the spray-type. After thorough study by health authorities, dishmachine manufacturers and others working with this phase of sanitation, minimum standards have been developed, published by the Joint Committee on Food Equipment Standards of the National Sanitation Foundation. These Standards have become guidelines acceptable to both health agencies and the industry.

1. Built-in or Under-Counter Type

(a) This is the smallest of the commercial dishwashers. It is used in nursing homes, small lunch counters, snack shops and for glasswashing at beverage bars. It is similar to home-type dishwashers, in that the door opens downward and the rack is pulled out onto the open door for loading and unloading.

(b) This machine is available as a free-standing unit with stainless steel top and side-panels or, without panels, to be built into a counter or bar.

(c) The wash and rinse waters are pumped through the same revolving wash arm. The tank is automatically drained after each wash or rinse cycle. The entire operation is controlled by an automatic timer.

(d) This machine can be equipped with a chemical sanitizing dispenser. *(See No. 4. Door-type chemical sanitizing machine.)*

2. Roll-type

(a) This is the smallest, above the table, spray-type model. It can be fitted into a small space, even in the food-serving area. The dishwasher (a waitress may be the operator in some cases) employed in an operation such as a small diner, tavern, luncheonette, or any quick-counter operation might well be using this machine.

(b) The roll-type machine has a rounded top which opens for loading. A wash water tank holds the detergent solution. After the operator has placed a rack of dishes in the machine and closed the top, he pushes the start button or control handle. The wash solution is then circulated in the required volume through a single revolving wash arm in the bottom of the machine. It is forced through the wash sprays up the sides of the curved lid and falls back, by gravity, over the dishes.

(c) After the washing period, a few seconds of pause or "dwell" are allowed to permit the wash water and detergent to drain off the dishes into the wash water.

(d) Last comes the rinse application which completes the machine's operational cycle. Rinsing is accomplished through wing-type sprayers above and a revolving rinse arm below. This 180°F. rinse, with its water flowing from a booster heater, will be covered in the discussion on sanitizing.

(e) The wash and final rinse cycles of this machine should be automatically timed, with provision for manual operation in case

Right: Under counter dishwasher.

Below: Door type dishwasher.

Corner installation of single rack door type machine.

of automatic control failure.

(f) Note that this is a stationary rack type machine—that is, the rack of dishes remains still while the wash solution moves around it.

3. Door-Type

(a) This machine is shaped like a box, with opposite doors sliding vertically upward for loading from either side. It may also be equipped with adjacent operating doors for a "corner" type operation. It, too, is classified as a single tank, stationary rack machine, but it has a greater capacity than the roll-type model.

(b) This is the model used in many large operations as a secondary unit for glasswashing. It is popular in those restaurant operations not requiring the load capacity of the large conveyor-type machines. It meets the needs of many school lunch programs, in-plant feeding installations, hospital isolation wards, etc.

(c) The door-type has a single tank for the water and detergent which are circulated in the required volumes and at the proper temperature by an electric motor-driven pump. The wash solution is pumped through the spray nozzles of two revolving wash arms, one above, the other below the rack of dishes.

(d) After a 3- to 5-second dwell period, a final cycle of fresh hot water rinses the dishes through a separate set of upper stationary and lower revolving rinse arms.

(e) Here again, the wash, dwell and rinse cycles should be fully automatic, with required time periods easily available for reference on the data plate. All such machines now being manufactured are required to have automatic time controls.

(f) In the event the machine has no automatic timer, the dishroom crew must be given some practical timing criterion, or the temptation to hurry through the operation during peak load periods will inevitably result in poorly washed, improperly rinsed dishware. For timing: The wash period should be at least as long as it takes the operator to prepare the next rack of dishware for washing, or 45 to 50 seconds. The first rinse should be operated for 12 to 15 seconds for each rack.

4. Door-Type Chemical Sanitizing Machine

(a) This machine may be found in small nursing homes and similar operations, or it may be used as a glasswasher.

(b) This, too, is a single tank, stationary rack model. The washing operation is similar to other door-types, but all wash water containing detergent is discharged to waste following each wash cycle.

(c) The final rinsing is accomplished with fresh water into which a sanitizing agent (a 5.2% sodium hypochlorite solution) is injected in sufficient quantity to provide the required sanitizing solution concentration. The operational cycle is automatically timed.

(d) Since this machine is designed to operate on building supply water temperatures (120° to 140°F.) and to sanitize effectively at these temperatures, a booster heater is not required.

(e) A simple chlorine test kit is available for frequent checking of sanitizing solution concentration.

(f) An iodophor sanitizing agent is also available for this type of machine, with an iodine test kit for frequent checking of sanitizing solution concentration.

(g) Silver plate or sterling silver should *not* be washed in this particular model since the sodium hypochlorite will cause tarnish. There will be no adverse effect on stainless steel flatware at the concentration used.

5. **Conveyor-Type Machines**

The larger models of dishmachines use a conveyor to move dishware through the wash and rinse cycles. They are of two general types.

(a) Rack conveyor-type—where the operator slides a loaded rack into the machine and the automatic conveyor carries the rack through the wash and rinse cycles.

(b) Rackless conveyors, on which about 70 percent of all dishes can be placed directly on the moving belt of the machine. Silver, glasses and cups can be more efficiently handled when sent through the machine in racks.

The rack conveying method may be either:

(c) An endless chain, running along each side of the machine's interior, which carries the rack on runners through the machine.

(d) A pawl drive with a reciprocating (moving backward and forward) action to pull the racks through the dishmachine.

(e) A continuous conveyor system having a dishwasher with an integral table track. Often built in oval, triangular or round configuration (structure), the layout can be adapted to meet the particular needs of the establishment for medium to high volume requirements.

Tableware is unloaded directly from bus box, cart or conveyor onto dishracks stationed on the table track.

As with many other types of equipment, the leading manufacturers of dishmachines have perfected their individual company models of the continuous conveyor system layout with resulting variations in method of garbage disposal, maintenance of clean table tracks, etc.

Soiled dishes are cleared over a trough at the side of the table track. Food soil or liquids in the trough may be funneled into some type of refuse disposer. In one model a table spray wash system cleans the table track during operation.

Soiled cups and glasses are placed in racks stored on overhead slanting shelf and put through machine as racks are filled. Silverware may be placed in racks, baskets or canisters. See further discussion under Racking, p. 91.

As clean ware emerges from the machine, it can be unloaded directly to adjacent shelving, self leveling dispensers or carts.

a. Single-Tank Rack Conveyors

(a) This model is often used in high school lunch programs, small institutions, in-plant feeding operations, medium-sized restaurants and other similar installations.

(b) A single tank contains the water and detergent at the required temperature. This wash solution is pumped through stationary or revolving upper and lower spray arms at the gallons per minute flow rate specified by health authorities and the machine's manufacturer.

Once the machine has been turned on, the wash solution flows constantly, whether or not a dish rack is in the machine.

(c) After passing through the wash spray pattern, the dish rack is then conveyed to the "clean" end of the machine where it activates a fresh 180°F. water final rinse application through spray nozzles located above and below the dishes.

(d) The speed of the conveyor is automatically timed for proper exposure of dishware to wash and rinse spray patterns.

(e) An interesting model of this type can be installed in a corner with expan-

Single-tank conveyor-type dishwasher.

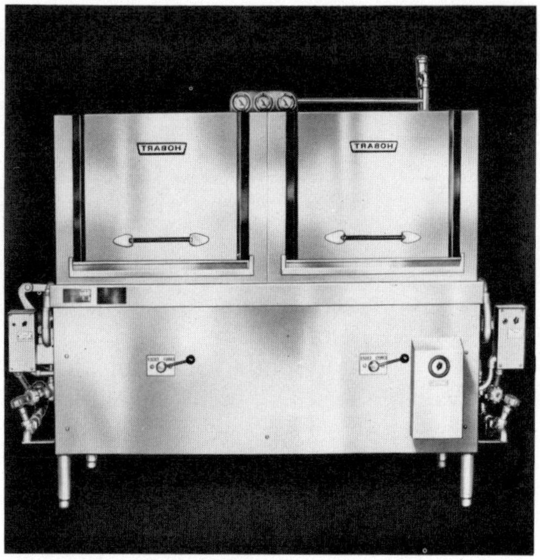

Two-tank conveyor-type dishwasher.

sion possible to convert it to a 2-tank conveyor type. After racks progress through the wash pattern, they can be moved around the corner automatically, then through the final rinse to the clean dish table.

(f) Many of these machines have had a power pre-wash unit installed ahead of the washing action, thereby achieving greater efficiency and economy.

b. Two-Tank and Multiple-Tank Rack Conveyors

(a) These models will, as expected, be found in the dish rooms of the larger foodservice operations. In addition to a power wash and a final rinse, a two-tank machine has a power rinse which re-uses the final rinse water, spraying it over the dishes to remove most of the detergent water and any remaining food soil.

(b) When a power pre-wash is built into this conveyor machine at the factory, with its own motor, tank and pump, it becomes an integral part of the machine and the model is sometimes referred to as a multiple-tank conveyor.

(c) In the wash tank the detergent solution flows through upper and lower solution stationary spray arms in the specified volume and at the required temperature and pressure.

(d) After the washing cycle, the rack is conveyed through a power rinse tank, where the dishes are rinsed free of most of the detergent solution by means of hot water pumped through stationary upper and lower spray arms, reusing water from the final rinse.

(e) The rack then passes through a final rinse, provided by hot water at sanitizing temperature, through upper and lower spray nozzles distributed evenly across the conveyor. An ample supply of hot water can be supplied through a booster heater.

(f) The conveyor is timed to carry the dishes through the series of wash and rinse cycles at a pre-determined rate of speed which, provided machine is properly operated and maintained, will result in the thorough washing, rinsing and effective sanitizing of dishware.

c. Power Prewash

(a) A conveyor machine may be factory equipped with a power prewash, having its own tank, pump and spray system. It is attached to, and is an integral part of, the dishmachine. It is similar in operation to the wash unit.

The power prewash tank may use spillover of wash tank water to maintain a changeover and water level at its own overflow. The spillover of wash water is created by the final rinse water from a pan which catches part of the used rinse water. In two-tank models, a by-pass line is used.

Another type of prewash system reuses water from the expended final rinse only, tempered with cold water and pumped under pressure to the prewash sprays.

(b) Depending on the motor horsepower and design of the prewash unit, it may remove as much as 90 percent of the gross food soil before the dishes enter the wash chamber, thus reducing the amount of detergent required in the wash tank to maintain proper detergent concentration level.

(c) The power prewash eliminates the need for pre-rinsing by hand, thus saving considerable labor costs. (Large pieces of leftover food should be removed by hand scrapping before racking the dishes.)

A power prewash may be installed on either a single tank or two-tank machine. A single tank machine with a power prewash added does not become a two-tank machine. A two-tank machine is properly described as one having a power wash, a power rinse and a final rinse, whether or not it has a prewash.

Many times dishmachine operators will speak of a machine in terms of performance, that is, what it can do. They will discuss a prewash, power wash, power rinse or final rinse. Thinking first of the functions or stages of activity inside the machine is a natural attitude. The quality of the dish crew's work is judged largely on the cleanliness of the tableware, the speed and efficiency with which it is delivered, and the appearance of the dishroom. To a great extent, the performance of the machine is the governing factor upon which the effectiveness of other crew activities depends. This is why the experienced operator loses no time in reporting any problem or sign of malfunction in the machine.

d. Rackless Conveyors

(a) Dishmachine personnel work with these machines in large hotels and restaurants, many hospitals or other types of heavy volume foodservice operations.

(b) Customer volume and dishroom space will affect the size of the model chosen. Rackless conveyors are available in single tank and two-tank models and models with a power pre-wash. Wash and rinse cycles of these three variously sized models so closely approximate those of the similarly sized rack-type conveyors that procedures do not need to be repeated.

(c) Trays are more easily handled when placed on edge on the conveyor, rather than by placing them in a rack, a consideration of importance in large cafeterias, etc.

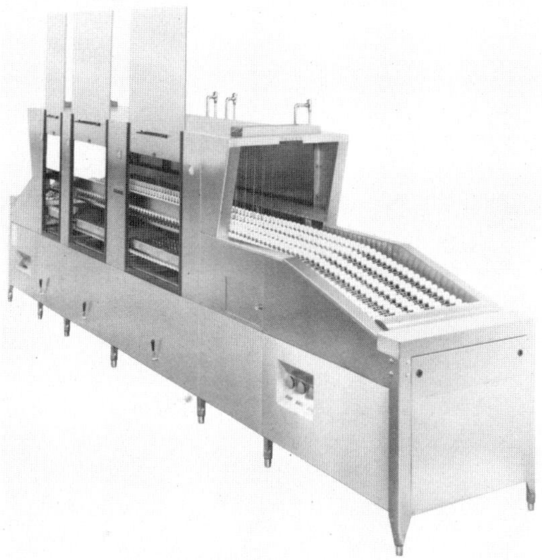

Rackless dishwasher, 2-tank unit with recirculating pre-wash.

(d) A switch at the clean end of the machine is activated when a rack or dish strikes it, stopping the conveyor. Thus, dishes are prevented from being broken against the end of the machine should unloading employees be too slow in removing them. Also, should a blockage occur in the machine, a slip-clutch is activated that prevents the conveyor motor from burning out.

(e) In all machines, the final rinse

Oval trays being loaded in large rackless conveyor machine.

line is activated only when the conveyor is moving. On some models, an electrically operated rinse gate may be installed so the rinse operates only when dishes are in the rinsing area.

Hot Water Temperatures

The hot water temperatures required for mechanical dishwashing may be placed in four categories. There may be slight variations in these temperature ranges, which we will discuss later.

1. **Pre-rinse water temperatures**

This water is used to remove as much food soil as possible from the dishes before they are washed. Suggested temperatures range from 110° to 130°F. depending on the method used to prewash.

2. **Wash water temperatures**

(a) This water, which combines with the detergent to form the wash solution, must be hot enough to remove all food soil from the dishes, dissolve grease from dish surfaces and activate detergent to its most effective performance relative to both cleaning ability and economy. Wash water temperatures are maintained by a heat source in the wash tank.

(b) The type and model of machine has a bearing on the required wash temperature.

For commercial spray-type machines, the National Sanitation Foundation's Standard No. 3 specifies the following wash water temperatures:

Stationary Rack Machines	150°F.
Single Tank Conveyor	160°F.
Multiple Tank Conveyor	150°F.

An exception is the door-type chemical sanitizing model, discussed under "Machine Types," which operates on the building's water supply at 120° to 140°F. temperatures.

3. **Power rinse**

The power rinse water temperature in multiple tank machines should be 160° to 170°F.

4. **Final rinse**

This final sanitizing rinse is specified as 180°F. before it leaves the rinse nozzles. This should insure a minimum sanitizing temperature of 170°F. on the dishware.

Dishwashing personnel are often puzzled by the stress laid on specified temperatures as well as by the variations occurring in different localities or even in establishments in the same locality. Perhaps the most simple explanation is that there are many ways in which heat can be lost during the dishwashing process and these temperatures are intended to compensate for such loss.

Dishes enter the machine at room temperature. Some heat is absorbed by dish racks, and by the interior and fittings of the machine itself. Every time a rack leaves a conveyor machine or a door-type is opened for removal of a rack, heat escapes. Therefore, the temperatures indicated on the various thermometers may not be the actual temperature of the dishware going through the wash-rinse cycles.

In order for the final sanitizing rinse to be an effective minimum of at least 170°F., the dishes must be many degrees hotter than they were when entering the machine and before they reach the final sanitizing rinse. The designated wash and rinse temperatures provide a measure of compensation and safety.

The required temperatures can be maintained by electric heat, gas heat or steam heat. Steam is used primarily with conveyor machines and is usually injected directly into the water in the wash and rinse tanks.

It is possible for the final rinse temperature to be *too* high. Temperatures above 200°F. tend to atomize a less forceful spray. At 210°F. no final rinsing is accomplished. Sanitizing cannot be accomplished under these circumstances.

Thermometers

Accurate indicating thermometers should be provided for the wash tank, recirculating power rinse tank and final rinse line. These thermometers should be checked frequently for accuracy of ±3°F. at the required temperatures.

A Dishwashing Operation

A two-tank, rack conveyor-type machine has been selected to illustrate dishwashing procedures. These procedures can be modified or expanded for application in any size or type of machine.

Large flight-type dishmachine in operation.

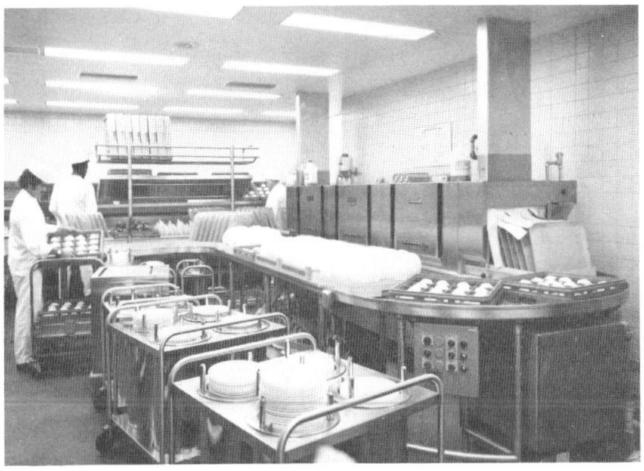

Large continuous rack conveyor machine. Note conveyor is L-shaped.

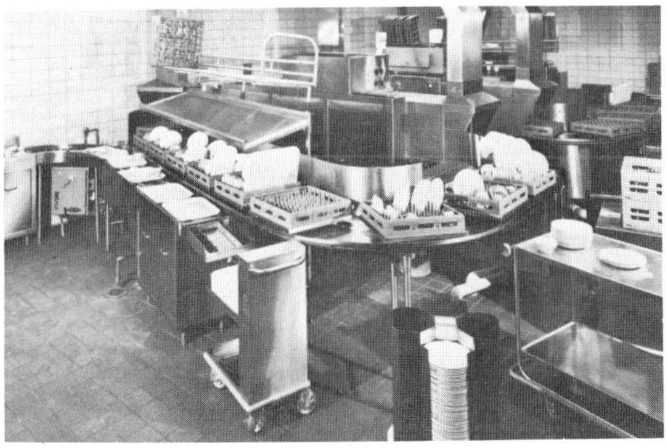

Continuous horizontal conveyor system in operation. Note plastic racks holding plates, trays, cups.

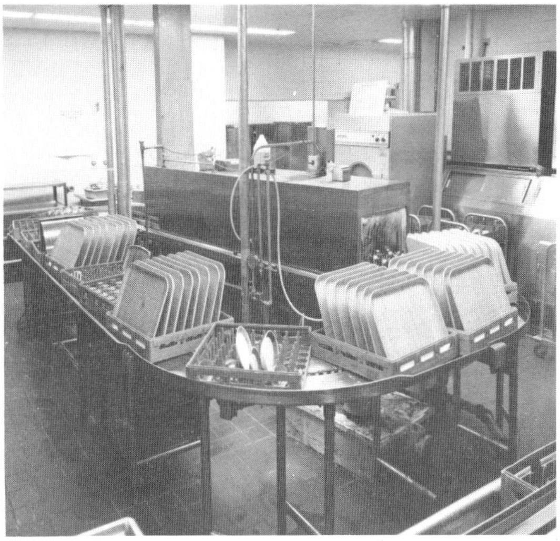

Continuous rack conveyor machine, sometimes called a "merry-go-round."

Pot and pan washing department, using machine. Note waste disposer, also clean pan storage area in right foreground.

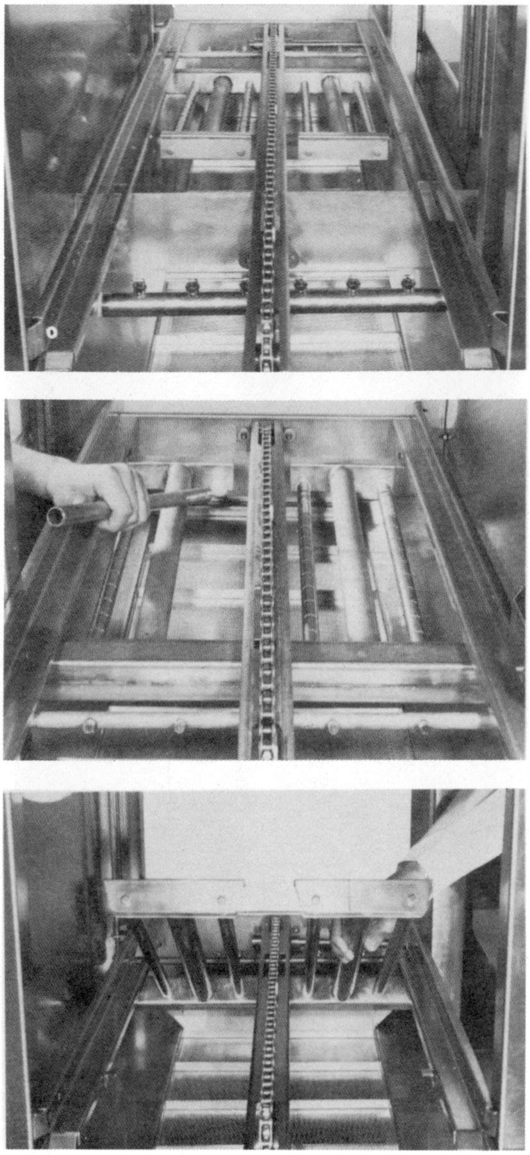

Dismantling of a dishmachine for cleaning of pipes, trays and interior.

Certain preliminary steps are to be carried out by dishwashing employees before the operation begins:

1. Recheck equipment to be sure it is ready for use:

(a) Inspect inside of machine. Remove any soil, broken glass, silverware or other foreign objects. BE SURE MACHINE IS CLEAN. Be sure overflow pipes are clean and free of debris.

(b) Inspect upper and lower spray arms. Be sure they are clean and properly installed. Check final rinse nozzles to be sure they are not clogged.

(c) Put all strainer pans and scrap baskets in place.

(d) Hang all curtains.

(e) Check detergent dispenser to be sure it is clean. Supply with sufficient detergent. Also check rinse injector equipment, inspect for cleanliness. Be sure it is ready for operation.

(f) Check entire dishmachine area; be sure dishtables and work areas are spotless.

2. To start machine:

(a) Shut drains.

(b) Open fill valve. Leave valve open till tanks are filled to top of overflow. **Note:** If machine has automatic or timed fill, turn "FILL" switch on.

(c) Add detergent according to method used. If machine has automatic detergent dispenser, turn dispenser switch on.

(d) Turn on tank heat and booster heater.

(e) Close inspection doors.

(f) Turn on motor switch.

(g) Run machine for one minute without dishes to allow for thorough combining of detergent and wash water at required temperatures.

(h) Check power wash and power rinse indicating thermometers for correct temperatures. Final rinse thermometer will indicate properly only when rack is in rinse compartment.

(i) Slide first rack of dishes into machine. Automatic conveyor will carry rack through to the other end.

(j) Slide rack into machine. DO NOT TRY TO FORCE RACKS THROUGH MACHINE FASTER THAN CONVEYOR MOVES THEM.

(k) Inspect first clean dishes carefully. If all parts of operation are apparently functioning properly but dishes still show soil or are insufficiently rinsed, report problem to supervisor. Continue thorough inspection, and reporting if necessary, throughout entire operation of dishmachine.

(l) It is advisable to allow a moment for clean tableware to air dry after coming through the machine.

(m) If machine has a "Drive-Dwell" control, turn lever to "Drive" to remove last rack from machine.

3. After dishes are done:
   (a) Turn off tank heaters.
   (b) Turn off booster heater.
   (c) Turn off detergent dispenser.
   (d) Turn off pump motor(s).
   (e) Clean the dish tables.
   (f) Open drain(s). BE SURE HEATERS ARE TURNED OFF FIRST.

4. Dishmachine cleaning will be discussed in a chapter of the same title, following "Sanitizing."

MACHINE DISHWASHING

1. **Scrapping**

(May be manual, hydraulic or both.)

Definition:

Scrapping is the removal from dishware of any pieces of solid or semi-solid foods or food refuse remaining after service.

(a) *Purpose:*

(i) To prevent the clogging of spray nozzles with food particles, thus assuring sprays will be directed in the designed manner. (A clogged spray may send water towards the side of the machine where it will accomplish no good.) Without proper scrapping, strainers and scrap trays, too, are less apt to become clogged with an accumulation of food residue.

(ii) To reduce the bacterial content of the wash water.

(iii) To keep the wash solution cleaner, thus lessening the need for too frequent changes of water and additions of detergent.

(iv) To remove as much excess soil as possible from the dishes prior to washing. Excessive soil may cause foaming and loss of spray pressure, requiring more frequent changes of water.

(v) To wet the soil remaining on dishes and thus make it easier to remove in the wash tank.

(b) *Methods:*

(i) Hand Scrapping—as previously described under Manual Dishwashing.

(ii) Water Scrapping—by use of flexible shower spray, either before or after racking. When used as a follow-up of hand scrapping, the power spray flushes off the small particles of food soil loosened but not removed during the hand scrapping operation.

(iii) Efficient equipment is available which provides a flushing action and automatic separation of the garbage.

(iv) Pre-wash units—where the dishes pass through a hydraulic scrapping department built in ahead of the dishwasher proper. The number of such built-in installations is increasing as the importance of thorough pre-washing is more and more emphasized. However it is done, the scrapping preparation of dishes for the actual washing is a procedure to be most carefully and thoroughly performed.

Plates, cups and other dishes should be properly racked to permit contact between washing solution and all surfaces.

2. **Racking**

(a) *Purpose:*

Scrapped dishes are placed in racks for conveyance through the dishmachine, or on a pegged conveyor for travel through it. The purpose of correct racking is to permit the wash solution and rinse water to contact *all* surfaces being cleaned.

(b) *Method:*

(i) Rack dishes openly, without overlapping. Remember that the backs of dishes must also be cleaned to prevent resoiling of food contact surfaces when the dishes are stacked.

(ii) Avoid trying to save time by crowding creamers or other miscellaneous pieces of dishware on top of dishes otherwise properly racked. Such jumbled arrangements result in ineffective washing, rejection of some of the rack's contents, and breakage.

(iii) Invert cups, bowls, any other deep dishes so that wash sprays can clean in-

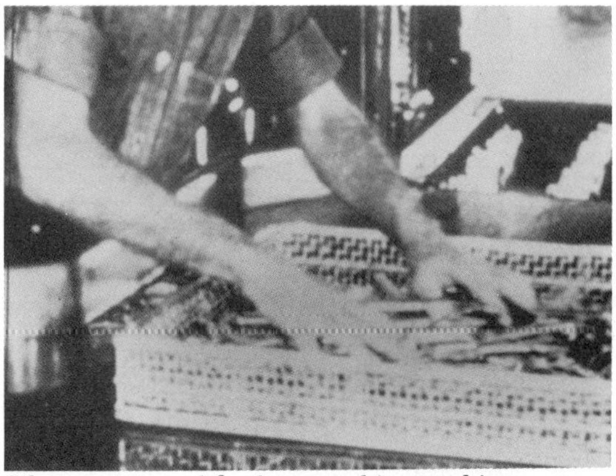

Silver positioned for best machine washing.

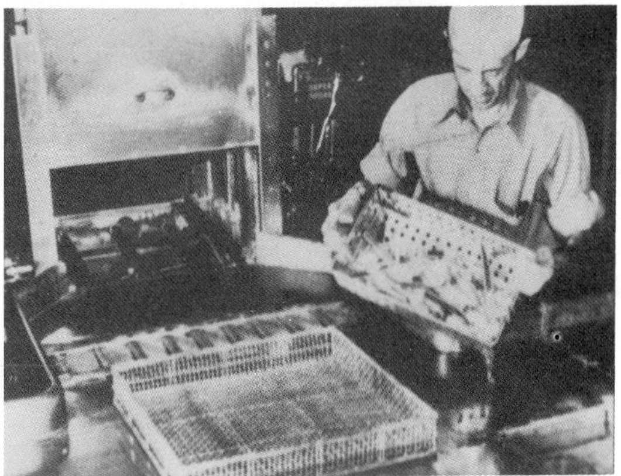

Dish racks that are overloaded are hard to handle.

While toweling of individual pieces of silverware is not recommended, clean towels are sometimes used to remove excess moisture prior to separating silverware. Hands should not touch silver. A small number of utensils should be toweled at a time, enabling worker to separate silverware without touching the bowls of spoons, tines of forks, blades of knives. Cylinders for holding silver make it easy to handle silver without touching the "business" ends. One cylinder is put through the machine empty, then silver is inverted when poured from full cylinder to clean empty cylinder.

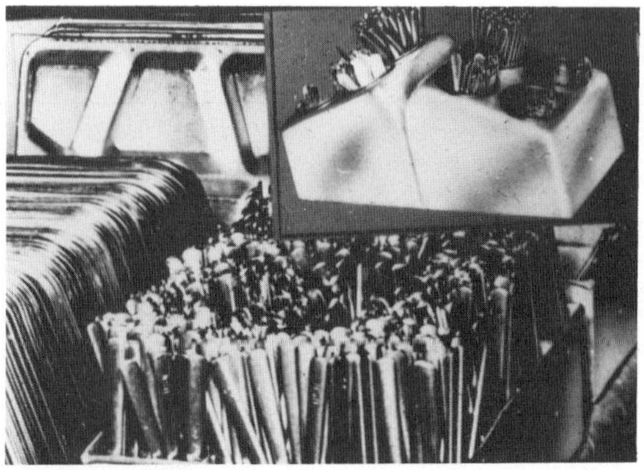

Sanitary storage and dispensing of silver. Note position of utensil handles.

ner surfaces without wash water being carried out into the rinse.

(iv) If flatware is washed in a conventional silver rack, pre-soaked pieces should be spread in a thin layer to insure effective washing and rinsing. Do not overload racks. Wash about 100 pieces of silver per rack.

Flatware is also sorted, placed handles down in perforated cylinders or compartmented, rectangular carriers. One empty silver canister goes through the machine on the same rack so that the washed silver can be inverted for service.

These silverware containers or baskets may be racked in open cup and bowl racks.

3. **Washing**

(a) *Purpose:*

To remove all food soil from dishes and dissolve grease on dishware. (The grease actually acts as a binder to hold soil on dishes.) Thus we have a transference of soil and grease from the dishware to the detergent solution where they are held until rinsing.

(b) *Method:*

As previously described, the detergent solution is circulated over and over again and keeps flowing through the upper and lower stationary or revolving spray arms at a controlled gallon per minute rate correct for this type machine. The correct water temperature is maintained by the heat source in the tank. The dishes are conveyed through the wash pattern at an automatically controlled speed. Fresh water from the booster heater line is constantly entering the wash pattern while excess water drains off through the overflow valve.

4. **Power Rinse**

(a) *Purposes:*

(i) To completely remove most of soil-laden detergent water from the dishware.

(ii) To allow a build-up of heat in the dishware, thus permitting desired effectiveness of the final sanitizing rinse and helping to ensure a quicker drying time.

(b) *Method:*

(i) After the dishes are passed from the washing area to the power rinse area, the hot pumped rinse water is sprayed over the dishes through stationary or revolving upper and lower spray arms in the correct volume and with a sufficient pressure to thoroughly rinse all portions of the dishes.

5. **Final Rinse**

(a) *Purpose:*

By the time the dishes reach the final rinse they should–if all previous parts of the operation have functioned correctly–be considered clean.

The purpose of the final rinse water's high temperature is to remove any remaining detergent and to inactivate bacteria remaining on the dishware after cleaning (including the disease-causing germs transmitted onto dishes and utensils).

All previous operational steps have led up to this final procedure. We must never underestimate its importance.

(b) *Method:*

The final rinse comes from hot water at sanitizing temperature, flowing in required volume and pressure through upper and lower spray nozzles distributed evenly across the conveyor. A drying agent is often injected into the final rinse to assist in a quicker, spot-free drying of dishes.

A discussion of the final rinse to assist proper unloading of dishes, further handling and storage will be covered in the section devoted to "Sanitizing."

6. **Pot and Pan Washing Machines**

Developments in the construction and cleaning procedures of mechanical potwashing machines are aimed at securing thoroughly clean pots and pans in the shortest possible time.

An excellent set-up for the potwasher. Note roomy drainboard on left for dirty pots, extra large soaking department. Next two compartments are for rinse and final rinse. Slanted draining shelf, open on left end, allows water to drain into washing sink below. Note bake sheet pan storage rack below clean pot drainboard, also the two free standing portable pot and pan storage racks opposite pan sink. Garbage disposer is located in lower left side of picture.

The preparation of pots and pans before actual washing is recommended with some models; in others no soaking or preflushing is said to be required.

In one such model, the principle of operation is to do one item of any size thoroughly and in less time. No preparation is needed if the soiled item had been used within the 2 hours preceding the washing.

If soiled ware is to be held for quite some hours or overnight, the items should be soaked. No other preparation will be needed.

Most of the items to be washed are held on a suction cup or base, powered by a small gearhead motor. A switch at the front of the machine releases the suction.

The wash water temperature is 140°F. The nozzle pattern of the wash spray is so structured that the largest volume of detergent solution is directed with high pressure through a rotating spray head, to contact the inside surfaces of the items positioned on the suction base. Secondary sprays are directed from the nozzles to contact outside surfaces of objects to be washed. Removable alternate racks are used for perforated pans or bowls or large sheet pans.

The wash cycles can vary from seconds to 5 minutes, depending on the type and amount of soil to be removed. A 10-second, 180°F. high pressure rinse follows the wash cycle. When some burned-on food is loosened, but not entirely removed, a light touch-up with a scouring pad will remove the remaining soil. The pot or pan is then put through a second wash of 10 seconds.

Study picture below right to review information in this section on dishmachine types and operations.

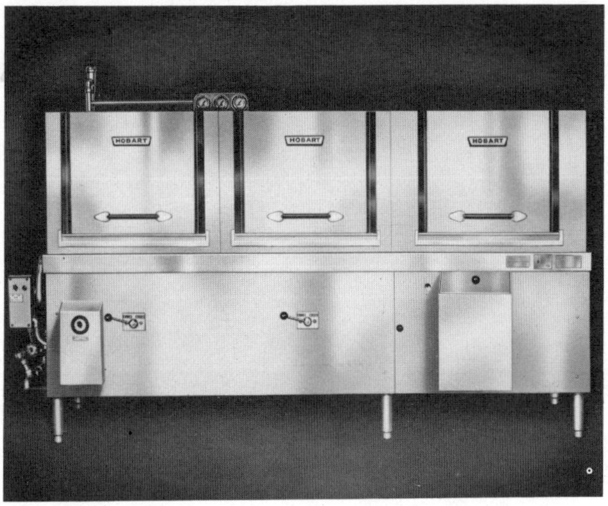

Two-tank conveyor-type dishwasher with 1 h.p. power pre-wash.

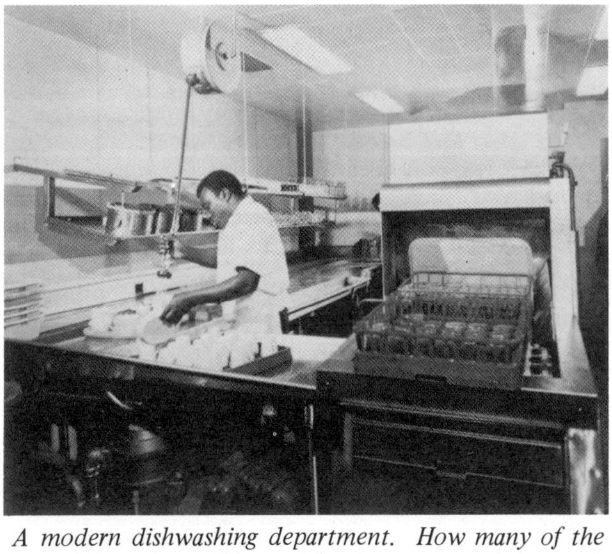

A modern dishwashing department. How many of the desirable features discussed in this chapter can you find? Check back through the chapter to see if you've found them all.

MECHANICAL DISHWASHING QUIZ

1. Fill in the following statements:
   (a) The basic requirements of ____________, _____________ and ______________ are common to both hand and machine dishwashing.
   (b) Dishmachine models are available for every type and size of foodservice operation. Name four of these models. ___________, ___________, ___________, ___________.

2. Check the correct statements:
   (a) The roll-type dishmachine is often used in luncheonettes. ( )
   (b) In the door-type machine, the rack of dishes remains still, while the wash solution moves around it. ( )
   (c) In conveyor-type machines, the dishware is moved through the wash and rinse cycles, either in racks or directly on the moving belt. ( )
   (d) 180°F. water is the *only* sanitizing agent used in machine dishwashing. ( )

3. True or False (check True or False):
   (a) Dishes enter the machine at room temperature. True ( ) False ( )
   (b) Thermometer on sanitizing rinse should read 170°F. while rinse is operating. True ( ) False ( )
   (c) Some of the heat escapes from a dishmachine every time a rack leaves a conveyor or a door is opened for removal of a rack. True ( ) False ( )
   (d) Racked dishes remain at room temperature in the machine until contacted by final rinse. True ( ) False ( )

4. Fill in the following statements:
   (a) Pre-rinse water removes as much _________ ________ as possible from dishes before they are washed.
   (b) Pre-rinse water temperatures can range from ____^{O}F. to ____^{O}F. depending on the method of pre-rinse or pre-wash.
   (c) Wash temperatures for commercial spray-type machines are:
       Stationary rack machine ___ ^{O}F.
       Single tank conveyor _____ ^{O}F.
       Multiple tank conveyor _____ ^{O}F.
   (d) The power rinse temperature in multiple tank machines should be a hot ____^{O}F. to ______^{O}F.
   (e) The final sanitizing rinse should be ________ ^{O}F. as the hot water leaves the rinse nozzles.

5. True or False (check True or False):
   (a) The higher the final rinse temperature, the better. True ( ) False ( )
   (b) When beginning wash operation, run dishmachine for 1 minute without dishes to allow detergent and water to combine thoroughly at required temperature. True ( ) False ( )
   (c) If first racks of washed dishes are clean, it is not necessary to inspect following racks. True ( ) False ( )
   (d) Scrapping will reduce bacterial count of wash water. True ( ) False ( )

6. Check the correct statements:
   (a) Washing should remove all food soil from dishes and dissolve grease on dishware. ( )
   (b) Wash and rinse time patterns should be automatically controlled. ( )
   (c) During a temporary lull, soiled towels or gloves can be run through the dishmachine. ( )
   (d) If dishes and glasses are washed in the same machine, save the glasses for the last. ( )

7. Fill in the following statements:
   (a) Dishes should be carefully racked without ________________________.
   (b) Cups, bowls, should be ________ to permit cleaning of inner surfaces without carry-over of wash water to rinse.
   (c) Silver in canisters should be placed with ___________ down.

8. True or False (check True or False):
   (a) During rush hours, wash as much silverware as possible in one rack. True ( ) False ( )
   (b) The purpose of the final sanitizing rinse is to destroy any bacteria remaining on the dishware after cleaning. True ( ) False ( )
   (c) If dishes are needed, it is OK to pull racks through the sanitizing rinse of conveyor type machines. True ( ) False ( )

9. Check the correct statements:
   (a) In conveyor type machines of rackless type, flatware, glasses and cups can be more efficiently handled when sent through the machine in racks. ( )
   (b) With prewashing, it is never necessary to change the wash solution in wash tank during a busy run. ( )

10. True or False (check True or False):
    (a) The dishmachine crew works with one of the most expensive pieces of equipment in the establishment. True ( ) False ( )
    (b) Dishwashing personnel help to protect food and protect health of customers and other personnel when dishwashing procedures are correctly carried out. True ( ) False ( )
    (c) The trained member of the dish ring crew is an important employee in the establishment. True ( ) False ( )

11 Sanitizing

Sanitizing is the final step in the dishwashing operation, the last act of the tableware and equipment cleaning routine. It is an especially important part of equipment and utensil cleaning, and it has a certain added meaning for dishwashing personnel.

The cleaning procedures leading up to sanitizing are carried out to provide the attractive looking tableware which management is proud to use and customers happy to see. Sanitizing is done to enhance this attractive appearance by protecting the health of everyone who uses the tableware while eating a meal in the establishment—management, personnel and customers alike. For these reasons, the dishwashing crews must regard the performance of correct sanitizing routines as one of their greatest responsibilities.

WHAT IS SANITIZING?

Sanitizing is a process or treatment carried out for the purpose of destroying bacteria (germs) remaining on the surfaces of equipment and utensils after washing and rinsing. In other words, sanitizing is applied to equipment and utensil surfaces which look clean and feel clean.

After removal of the food soil which can be seen and felt, dishwashing personnel have the all-important task of removing the soil which cannot be seen or felt.

There are several ways in which tableware and the food-contact surfaces of equipment may be sanitized. The routine selected, however, must be one approved by the health authority concerned. Health department regulations state that the process used must have been proved successful in destroying germ life, including disease-causing bacteria. When a sanitarian makes an inspection of the establishment, therefore, he never fails to check the sanitizing method and test the results.

Terms Used in Discussing Sanitization

In public health work, a number of different terms are used in connection with the methods and materials employed to destroy pathogens (disease-causing organisms). These terms, or words, though not exactly the same in meaning are often used interchangeably, that is, they may be substituted for each other as though their meanings were exactly the same.

Foodservice personnel may often become confused and misled by hearing many different terms used to describe what they believe to be the same type of product or the same procedure. A brief discussion of these most commonly used words will help employees to understand their differences in meaning.

1. First, we will study the word *disinfectant,* because there are thousands of registered brands of disinfectants sold in the United States. These disinfectants are of many different types but intended for the same general purpose, the destruction of infection-producing organisms. To understand the word disinfectant, we must understand what infection is and how infections are passed along.

What is an *infection?* Simply a condition where the body is in an unhealthy state because some disease-causing germs, bacteria, viruses, etc. have gained entrance to the body, reproduced themselves and multiplied to the point where our good health is affected by their presence and actions.

COMMON TYPES OF INFECTIONS–A boil is one type of infection. Many of us have had *infected* cuts or sores. Respiratory infections, which many people call heavy chest colds, are common. There are also certain infectious diseases which put special focus on the work habits of personnel because these diseases can be passed along from person to person in food establishments.

PASSING INFECTIONS ON–One way in which this can be done is through the use of dishes, utensils and equipment on which disease germs are living. When food or drink is placed in contact with such surfaces, the germs are transferred from the food to the mouths of the people who eat it, and illness may well result.

The Communicable Disease, Food Infection, and Food Poisoning sections cover food infections and communicable diseases in quite some detail. What we are doing here is becoming familiar with the word *infection* so that we can better understand *disinfection* and *disinfectant.*

A *disinfectant* is a substance, usually chemical, which destroys infection (disease causing microorganisms).

A microorganism is a very small form of plant or animal life, so tiny it cannot be seen without using a microscope. Bacteria are microorganisms.

Some bacteria form hard walls or shells around themselves. These forms are called spores. These spores resist the killing action of many disinfectants and are not always destroyed by them.

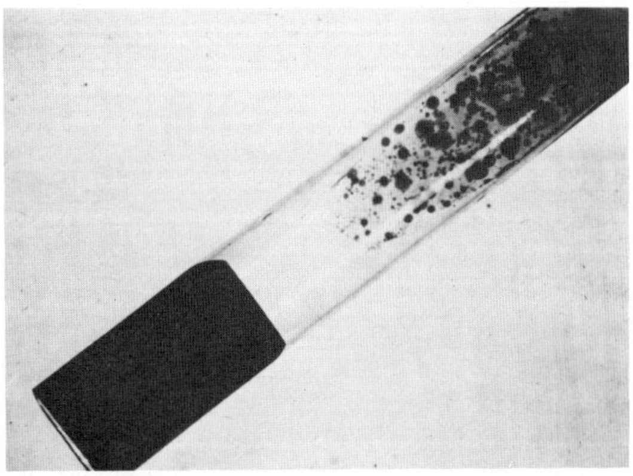

This tube holds colonies of bacteria growing on sterile food. These bacteria grew in a sample of water taken after 700 utensils had been washed in it. No sanitizing agent had been used in the water so bacteria remained in the water. This slide was produced by the Rockford, Ill. health department.

Disinfectants are applied chiefly to inanimate (not living) objects rather than to living tissues. Thus, we disinfect rooms, furniture, mattresses, clothing, to destroy disease germs present in these areas or on these articles. It is also correct to speak of the *disinfection* of dishes and utensils, since these are inanimate objects.

On the other hand, there are many products in the disinfectant class which would be most unsuitable for use on food establishment tableware, utensils and equipment. So, to avoid confusion, a different term is used in foodservice to describe the bacteria-destroying treatment applied to food establishment eating and drinking utensils and the food-contact surfaces of equipment.

2. The word used in foodservice is *sanitize.* A sanitizer is an agent, usually chemical, which is applied in a separate operation after the objects to be sanitized have first been cleaned. In food and/or drink establishment sanitation regulations, the approved sanitizing agents are not required to destroy all of the microorganisms present on eating and drinking utensils and equipment. But they must reduce the number of these organisms to the safe level acceptable to health departments.

Sanitizers destroy both disease-causing bacteria and those regarded as harmless.

Sanitizers may be combined with detergents to form products called detergent-sanitizers, detergent-germicides, detergent-disinfectants or some other equally descriptive term, the name depending largely on the chemical formulations of the products. These will be discussed in more detail under "Sanitizing Methods."

3. *Germicides.* A germicide is an agent which destroys microorganisms. But the term refers in particular to the chemical agents which destroy disease germs. Germs are thought of more often in reference to disease-producing organisms, while bacteria may be disease-causing, helpful, or harmless. Some germicides are formulated for use on inanimate objects while others

may be applied to living tissue.

4. A *bactericide* is an agent which destroys bacteria, including both the disease-causing and the non-dangerous types. The term is applied especially to chemical agents used for the above mentioned purpose. *Bactericidal treatment* is the application of a bactericide, in accordance with public health regulations, to objects such as food-service equipment, dishes or utensils. Hence, when effective bactericidal treatment has been applied to such objects, they have been correctly sanitized.

5. *Sterilize.* This is a process, physical or chemical, which destroys all forms of life, including microorganisms. A sterilizing treatment, at the correct temperature and time exposure, will stop the activity of viruses and spores.

Many personnel will associate the term *sterilize* with hospitals, operating rooms, surgeons' instruments, dressings, hypodermic needles, syringes, wherever a situation demands that individuals or equipment be made completely free of any possibility of infection.

Sanitization is not sterilization. Therefore, it is not correct for food establishment dishwashing personnel to believe or state that they are sterilizing their dishes, utensils, or equipment.

6. *Antiseptic* is another term which many personnel do not understand. An antiseptic is an agent which is applied to living tissues, such as open sores, cuts, wounds, for the purpose of destroying microorganisms or stopping their growth in action on the tissues.

Familiar antiseptics such as tincture of iodine, tincture of metaphen, mercurochrome, and similar products will be found in most home medicine cabinets.

7. The meaning of pH will be discussed briefly for two reasons:

(a) Dishroom supervisors, dishwashing employees or others who have occasion to read the labels, descriptive literature, or use instructions for detergents, detergent-sanitizers, or sanitizers will undoubtedly find references to pH. Employees will find it helpful to understand, in a general way, what such references mean.

(b) Many terms are self-descriptive, pH does not explain itself to the casual reader, neither does it provide the dishwashing crew with any clue to its meaning.

The letters "pH" followed by a number show whether the detergent solution is alkaline or acid in content and to what degree it falls into either of these classifications. In other words, pH tells the strength of the alkalinity or acidity of the solution.

Why does the alkalinity or acidity matter to the employee? Personnel are very familiar with both types of solution.

Alkaline or acid compounds are often combined with other chemicals in the manufacture of cleaning products. The alkalis, in particular, are probably the oldest and most commonly used chemicals found in industrial cleaners, including the dishwashing compounds.

The acid type solutions have more specialized uses, such as deliming dishmachines, removing water spots or film buildup from dishes, etc. Refer to our detergent section for a discussion of detergent ingredients, combined to create balanced chemical cleaners for specific purposes.

The alkaline or acid content (pH) of a detergent solution is not only an indication of how the solution should be employed, but is a guideline to its most effective usage.

Imagine a ruler marked off into units or sections from zero to 14. The number 7 is in the middle of this ruler, or scale.

The imaginary ruler does not measure inches, feet, etc. This is how the alkaline or acid strength of a detergent solution is shown.

The number 7, being in the exact middle of the scale, is the point where neither acid nor alkali is the stronger.

Each descending unit below 7 (6, 5, 4, 3, 2, 1) stands for an acid strength 10 times as great as the unit just above it. The lower the number under 7, the greater the acid strength in the solution.

Returning to the number 7 and proceeding higher on the scale (8, 9, 10, 11, 12, 13, 14) each ascending unit *above* 7 stands for an alkaline strength 10 times as great as the unit below it. The higher the number above 7, the greater the alkaline strength of the solution.

In a manual dishwashing operation,

where the hands are in direct contact with the detergent solution, a pH range on the scale from about 4.5 to 9 or 10 shows that the product, at its indicated strength, will not irritate the skin, and is considered safe for hand use. Individuals, especially if they have tender skin or bruises on their hands, may question the accuracy of these pH figures. Workers who are prone to contract skin problems should, perhaps, seek some other area of employment rather than in the dish room.

Dishwashing compounds or other cleaning products high in alkalinity or acidity would be unsuitable for hand dishwashing or cleaning operations but would, for instance, be effective in the dishmachine.

Thus, we see that pH helps the manufacturer to provide the proper solution strength for the intended usage of his product, as well as indicating its uses in the field. We have seen how this works by noting the difference in pH between compounds designed for hand and mechanical cleaning operations. We will find additional references to pH in chemical sanitization.

SANITIZING WITH HOT WATER IN MANUAL DISH AND POT WASHING

At the conclusion of the Manual Dishwashing discussion, the dishes had been properly washed in Sink No. 1 and thoroughly rinsed in Sink No. 2. (The 100°F. warm rinse water was kept clean by allowing it to run continuously and overflow.) The dishes, clean to sight and touch, were left ready for sanitizing in Sink No. 3.

When hot water is the sanitizing agent, a water temperature of at least 170°F. must be maintained in Sink No. 3 in order to destroy bacteria. Eating and drinking utensils and equipment items must be immersed (placed completely beneath the surface of the water) in the clean 170°F. water for at least one-half minute.

These sanitizing requirements mean that the following items must be made available:

1. A sufficient amount of clean hot water at the 170°F. temperature to keep the sanitizing water in Sink No. 3 at its proper heat level all during the dishwashing period. This situation will be handled by

Keep final rinse water in dishmachine at 180°F. This provides at least 170°F. at all points of contact with dishes, cups, etc.

management, but any drop in temperature of water flowing into the sink during the sanitizing treatment should be reported immediately to the dishcrew's supervisor.

2. A wall thermometer, mounted over the No. 3 sink, so that dishwashing personnel may check the water temperature at any time. This thermometer must be accurate to within ± 2°F. Use of a wall type thermometer will reduce the costs of breakage or loss of these instruments.

Frequent reference to the thermometer will be necessary. The temperature of the dishes and utensils placed in the hot water bath will be under 170°F. and will, therefore, lower the temperature of the sanitizing bath. Thus, the water flowing into the sink should be at least 180°F. to allow for this drop in temperature.

3. Dish baskets with long handles, designed so that the dishes, utensils and small pieces of equipment placed in the baskets may be completely covered by the sanitizing water.

The long handles on the baskets permit dishwashing personnel to carry on the sanitizing treatment without their hands coming in contact with the 170°F. water.

4. Tongs or hooks for the handling of large or heavy equipment which can be correctly sanitized in Sink No. 3, but cannot conveniently fit into a long handled basket.

5. A clock with a second hand should be located within convenient view of the dishwashing crew. It takes time to destroy bacteria. Dishes and utensils may remain in

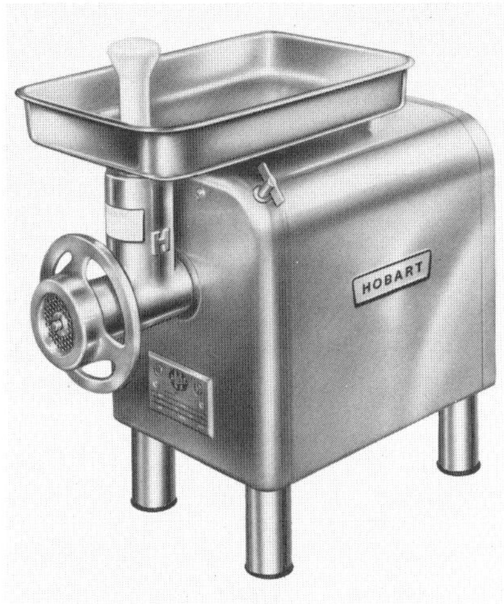

Slicers and other food equipment also needs to be cleaned and sanitized.

Proper sanitization of large utensils is also important. Hooks or tongs may be needed to handle such items.

the 170°F. water longer than one-half minute, but not even one second less if sanitization is to be accomplished. This minimum time element is very important.

When transferring washed dishes to the long handled baskets, care should be taken in the racking of cups, bowls, etc. If air is trapped in a dish of this shape it will prevent the hot water from contacting all surfaces of the dish and it will not be sanitized.

Dishwashing personnel who wash pots, pans and other food-contact equipment surfaces have sometimes asked if it is really necessary to sanitize these equipment items. The simplest answer is that some germs concern themselves with attacks on food, and they gain entrance to the food when it has contact with unclean, insanitized equipment.

It is easy to detect the presence of some bacteria. They make certain foods sour, particularly milk, cream and foods which combine these dairy products in their preparation. All personnel are familiar with this type of spoilage.

On the other hand, insanitized equipment and/or utensils, though visibly clean, may have bacteria remaining on their surfaces which are capable of infecting food, causing food infections or food poisonings.

Where equipment, even when disassembled, is too large to be sanitized in the sink, it may be satisfactorily treated as follows:

(a) By rinsing carefully with boiling water or

(b) With live steam from a hose, in the case of equipment in which steam can be confined.

After being immersed in 170°F. water for at least one-half minute, the loaded dish baskets should be placed on a clean, slanting drainboard while their contents air dry. Under no circumstances should a towel be used to dry dishes or utensils. It is impossible to keep towels clean or free from hand contamination. If used, the effect of the sanitizing treatment is immediately destroyed.

The hotter the sanitizing water, the faster the drying and the less chance of water spotting. In some areas the drying process is hastened by the addition of a small amount of rinse additive to the sanitizing water. The local health department should be consulted before this is done.

Personnel must use the greatest care in removing sanitized dishes, utensils and equipment items from the long handled baskets. Sanitized spoons, knives and forks are to be picked up or touched only by their handles. Sanitized cups, glasses and bowls are to be handled so that fingers do not contact inside surfaces or lip-contact surfaces. Plates are to be handled by their rims. The same precautions should be followed when handling the food-contact surfaces of the equipment items small enough for hot water sanitization in Sink No. 3.

Proper storage will be discussed after completion of the sanitization section.

There have been many situations where an establishment's hot water has been insufficient in volume, temperature or both. Sometimes other water supply lines, in drawing water off the main lines, have seriously lessened the amount of hot water needed for the dishwashing operation. In other cases, personnel have been careless in following correct routines by disregarding both the time and temperature requirements for hot water sanitization. Some dishwashing employees dislike working with 170°F. water, even when provided with proper tools.

Sanitizing with approved chemicals is also carried on in large numbers of manual operations. Hot water and chemical sanitization are equally effective if properly applied.

SANITIZING WITH APPROVED CHEMICALS IN MANUAL OPERATIONS

We return again to the 3-compartment sink in which the washed and rinsed dishes were left ready for sanitizing in Sink No. 3. This time the sanitizing will be done in an approved chemical solution.

General Requirements

1. Except where specified in local food sanitation regulations, the 3-compartment sink is a standard requirement for a routine wash, rinse, and chemical sanitizing operation. The clean, running, and constantly overflowing rinse water in sink No. 2 will remove any traces of loosened food soil and detergent solution remaining on the dishware after washing. The chemical and water solution in Sink No. 3 can then do an effective job of sanitizing.

2. The fact that chemical sanitizers do not work satisfactorily on unclean surfaces is important for dishwashing personnel to remember. It re-emphasizes the necessity for thorough washing and rinsing of the ware prior to sanitizing. In addition, the wash, rinse, and sanitizing sinks must be kept clean as well as the tools and other equipment (brushes, long handled baskets) used in the wash-rinse-sanitize routine.

3. To avoid recontamination by improper handling, the use of long handled baskets is recommended during the rinsing and chemical sanitizing.

4. To make up the correct sanitizing solution, the exact amount of water in the sanitizing sink must be measured. If the manufacturer's instructions state that 1 tablespoon of a sanitizer is to be added for each 2 gallons of water, it is obvious that the total number of gallons of water in the sink must be known before the proper dosage of sanitizer can be added.

Health department ordinances regulating food and/or drink establishments generally express the required amount of sanitizer in the solution as so many parts per million (ppm). Manufacturers' use instructions, however, usually translate this requirement into specified numbers of tablespoons or ounces to stated amounts of water.

5. The effectiveness of the sanitizer gradually becomes less as it contacts more and more dish and equipment surfaces. Hence, the solution should be changed or renewed whenever it falls below the required strength. For solution testing purposes, inexpensive, easy-to-use test kits may be purchased or provided by the manufacturer. It is advisable to test the solution strength after sanitizing about 400 articles.

6. The strength of the sanitizing solution and the immersion time must meet requirements in the U. S. Public Health Service Food Service Sanitation Manual. In these examples, the immersion times given are the minimum or shortest time period allowed for effective chemical sanitization (see facing page). Various state and local ordinances may expand these requirements.

The final authority in each area is the local health department and the code prepared by that department. Foodservice operations should always follow the rules specified by their local health officials.

Chemicals Used in Manual Sanitizing

1. Chlorine is undoubtedly the most familiar and probably the most widely used chemical for sanitizing purposes. It was a popular sanitizing agent as long ago as 1900. Chlorine is available in liquid or powder form; manufacturers' use instructions are specific for each type.

2. Iodine, long a well known germicide, is approved as an effective sanitizer for eating and drinking dishes and utensils, and is non-staining and non-irritating in the concentrations required for sanitizing.

3. Quaternary Ammonium Compounds are also available as third sink sanitizers. Manufacturers point out that the requirements of local health codes must be followed for use of this type product.

Chemical Sanitizing Procedures

USPHS Food Service Sanitation Manual 1962, page 54.

"(b) Immersion for a period of at least 1 minute in a sanitizing solution containing:

"i. At least 50 ppm of available chlorine at a temperature not less than 75°F.; or

"ii. At least 12.5 ppm of available iodine in a solution having a pH not higher than 5.0 and a temperature of not less than 75°F.; or

"iii. Any other chemical-sanitizing agent which has been demonstrated to the satisfaction of the health authority to be effective and nontoxic under use conditions, and for which a suitable field test is available. Such sanitizing agents, in use solutions, shall provide the equivalent bactericidal effect of a solution containing 50 ppm of available chlorine at a temperature not less than 75°F.

"(Note) It is recommended that, when made up, the strength of sanitizing solutions be at least twice the minimum strength required for the particular sanitizing solution used."

The Manual also notes that if equipment is too large to be chemically sanitized in a sink, as explained above, it may be treated by spraying or swabbing with a chemical sanitizing solution of at least twice the minimum strength required for the particular sanitizing solution when used for immersion sanitization. The chemical sanitizer is effectively applied to equipment surfaces in the form of a fine spray or by a fogging treatment; treated area should be tightly closed to contain the fog. Special spray nozzles and fogging apparatus have been perfected for these types of treatments.

While it is interesting and helpful for students or employees to know something about the chemicals used in sanitizing tableware and equipment, the most important items to remember are:

1. Manufacturers' use instructions must be understood and followed to the letter.

2. This means that measurements of chemical and water must be exact.

3. Solution temperatures must be as required.

4. After chemical sanitizing, dishes and utensils should be allowed to air dry in baskets or racks on a clean, self-draining (slanting) drainboard. Since solution temperatures are lower than required for hot water sanitization, the drying time will be somewhat longer. Hence, good ventilation of the dishwashing area is a "must" and drainboards should be sizable enough to accommodate several baskets at the same time, if necessary. The sanitized ware should not be towelled dry.

Hot Water as Sanitizing Agent

In the majority of dishmachines, hot water is the sanitizing agent. The hot water may be the fresh water rinse, as in door type machines; or the final rinse, as in conveyor type machines.

The temperature of the final sanitizing rinse, sprayed at required volume and pressure, should be at least 180°F. as it leaves the rinse nozzles. If all other factors in the dishwashing operation are being correctly performed, the 180°F. hot water will provide effective sanitization of the items being rinsed.

Recalling again that the temperatures of dish and utensil surfaces are considerably less than 180°F. after the washing process, we can understand the necessity for the minimum of 180°F. for the sanitizing rinse. This temperature will raise the tableware surfaces to the degree of heat needed to reduce bacterial populations on the ware to acceptable levels.

While it is important that the final rinse be at least 180°F. it should not exceed 195° to 200°F. Water at these higher temperatures, when released from pressure in the rinse line, will atomize and become vapor, with the result that the necessary heat impact of a liquid rinse is lost before it can contact dish and utensil surfaces.

An easily readable, enclosed, sealed type thermometer, accurate to within $\pm$ 2°F. should be installed on the dishmachine to show the temperature of the sanitizing rinse water as it enters the spray arm or manifold of the final rinse.

The dishmachine lesson dealt with various models of spray-type machines. Our sanitizing discussion will follow this same pattern. At this point, personnel may wish to refresh their memories on the model, construction, etc. of the machine with which they are working.

1. STATIONARY-RACK: HOOD AND DOOR TYPES

The wash and rinse cycles of all machines of this type should be automatically timed. If, however, manual operation becomes necessary, the minimum rinsing time is 10 seconds, during which time each rack of dishes is uniformly sprayed with hot rinse water of 180°F. water temperature at the entrance to the rinse manifold.

There are also water volume and flow pressure requirements which, as mentioned in the dishwashing lesson, will not be detailed for personnel. If a situation arises where the dishes are poorly rinsed despite clean wash and rinse arms, spray jets free of debris, correct hot water temperatures and rinsing time, a supervisor should be notified immediately. A trained worker will then check and correct any defect in water volume or pressure.

2. SINGLE-TANK CONVEYOR TYPE

When the conveyor moves a dish rack to the *clean* end of the machine, a hot water sanitizing rinse is activated and released through spray nozzles located above and below the dishes and distributed evenly across the conveyor. The speed of the conveyor is automatically timed for proper exposure of dishware to wash and rinse spray patterns. The conveyor speed of this machine is set at the factory and cannot be changed. Personnel are warned against pulling the racks from the conveyor before their contents have been exposed to the sanitizing rinse for the proper length of time.

3. MULTIPLE-TANK CONVEYOR MACHINES

The final 180°F. sanitizing rinse is provided through upper and lower spray nozzles distributed evenly across the conveyor. The rinse volume is based on a rack or conveyor width of 20 inches. The wider the unit, the greater the rinse volume required.

The conveyor is timed to carry the dish racks through the series of wash and rinse cycles at a pre-determined rate of speed which, provided the machine is properly operated and maintained, will result in the thorough washing, rinsing and effective sanitizing of tableware. The conveyor speed of this machine is set at the factory. This type machine was used in the dishwashing lesson to illustrate a correct dishmachine operation.

In many establishments it is the custom for dishwashing employees who unload the machine to protect their hands from the heat of the dishes by wearing white gloves. Often these employees are under the impression that white gloves also insure a sanitary handling of tableware.

It is true that visible dirt will probably be noticed sooner on a white glove than on a bare hand, indicating that the soiled glove should be replaced with a clean one. This, in itself, could be considered a help in establishing a sanitary environment.

It is also true, however, that a white gloved hand may wipe a face, scratch a head, rub a nose, or engage in other undesirable hands habits. The white glove can then become just as contaminated by invisible soil as a bare hand, or a hand towel would be under the same circumstances.

Chemical Sanitizing in Dishmachines

This single-tank, stationary rack door-type machine operates on building supply water temperatures (120° to 140°F.) and effectively sanitizes at such temperatures. After the wash solution is discharged to waste, the final rinsing is accomplished with fresh water into which a sanitizing agent (a sodium hypochlorite or iodophor solution) is injected in sufficient quantity to provide the required sanitizing solution strength. The operational cycle is automatically timed.

The strength of the sanitizing solution may be easily and frequently checked by an available chlorine test kit or an IODE test kit.

Detergent Sanitizers

A detergent-sanitizer is a balanced chemical cleaning compound which has had a sanitizing agent added to its formulation. Products in this group may be called germicidal detergents, detergent disinfectants or other descriptive name, depending to a large extent upon the type and special qualities possessed by the sanitizing agents in the compound.

To be effective, the sanitizer should have rapid action and thus must be combined with cleaning ingredients which will not lessen or slow down the killing ability of the sanitizing chemical. Thus, since certain sanitizers work best in acid solutions while others prefer alkaline type compounds of varying strengths, a correct pH becomes an important factor.

Chlorinated compounds were described under "Detergents." These have been used for stain control of dishware, glass washing, the lowering of the bacterial population of the wash solution, the general freshening of the dishmachine's interior, etc. as well as for other food establishment cleaning routines.

The sanitizing effect of chlorine is rapidly lost in the presence of soil. The higher the alkaline content of the washing solution, the less effective the chlorine becomes.

Iodine has also been combined with one or more suitable water softening agents plus,, frequently, phosphoric acid. This combination of chemicals produces the detergent-sanitizers known as iodophors. When used as directed, these compounds are nonstaining and have been recommended in hand dishwashing operations.

Where two-compartment sinks are installed on the premises, these compounds have been used to lower the bacterial population of the wash water. Many health departments, however, require that their use be followed by a sanitizing rinse. Check the local food ordinance for specific usages.

Quaternary ammonium compounds are also combined with other chemicals to create detergent-sanitizers. The quaternaries act more effectively in alkaline solutions than in the acid types. They work better as the pH is increased. Attention is often drawn to the quaternaries' deodorant abilities. Aside from their other suggested uses in hand dishwashing, they are frequently applied in areas such as toilet and garbage rooms where odor control is an important factor.

Dishwashing and maintenance crews need not concern themselves too much with the chemical make-up of detergent-sanitizers, germicidal detergents, etc. The examples listed above, with their brief and incomplete use references, have been provided mainly to let personnel realize they are dealing with a special, complex and very interesting type of product.

Whether these products are used in dishwashing, floor mopping, pressure spraying of equipment or other applications, there are a few general rules which should be followed to ensure the most satisfactory results. These would include:

1. Follow the manufacturers' instructions to the letter. Measurements must be exact. The manufacturer generally requires a certain specific amount of the compound to a stated amount of water, depending here on the intended usage of the product. The amount may be given in tablespoons, ounces, per gallon or more of water. In hand dishwashing operations, this means that the exact amount of water in the wash sink must be known before the detergent-sanitizer is added. The same careful measuring should be observed in dishmachines when chlorinated compounds are used. Since the majority of modern dishmachine operations include some type of detergent dispenser, dishwashing personnel working on machines are helped with their compound measuring. Manual operators, however, must learn to be exact and conscientious about adding the required amount of detergent-sanitizer to the wash water.

2. The importance of keeping any wash solution as clean as possible by frequent changes, hand scrapping, pre-rinsing, pre-washing, has been stressed over and over again. This is especially important when detergent-sanitizers are used.

We already know that a sanitizing treatment cannot work effectively unless the dishes, utensils and equipment are free from soil. In a detergent-sanitizer and water solution, therefore, the rapid action of the sanitizer will become both slower and less effective as more and more soil builds up in the solution. Again, we have particular reference to hand dishwashing operations.

For this reason, dishwashing personnel should inspect their wash solution even more often than they do when their compound contains no sanitizer. They cannot permit the wash solution to become soiled to the point where the sanitizer in the compound cannot perform its work of destroying bacteria.

3. Correct washing temperatures must be strictly observed. The instructions for the use of certain detergent-sanitizers, for instance, specify "Do not use in hot water." Lukewarm solutions from 75°F. to 110° or 120°F. are manufacturers' instructions for others.

4. The 3-compartment sink operation, with the No. 2 sink water used as a rinse prior to sanitizing in sink No. 3 is required in most food and drink establishment sanitation ordinances. In this set-up, the use of a detergent-sanitizer in the wash solution, followed by a clean water rinse, then the immersion in the sanitizing solution in sink No. 3 has resulted in effective sanitization of dishware. But any variations from the 3-compartment sink set-up should always be avoided until health authorities have been consulted.

Over recent years many new warewashing products for manual or machine use have been developed to meet ecology requirements. A considerable number of the new detergents, in both powder and liquid form, contain a sanitizing agent such as chlorine or iodophors.

Correct and incorrect way to serve a glass. Sanitizing efforts have been wasted when glass is picked up by the rim.

STORAGE OF DISHES, UTENSILS, EQUIPMENT

The necessity for proper handling and storage of cleaned and sanitized dishes, utensils and equipment requires careful thought by all food establishment personnel. The main facts to be considered are these:

1. Money, time, and effort are spent by management and dishwashing personnel in

(a) Providing the necessary equipment and tools for a sanitary dishwashing operation;

(b) Learning why and how certain work routines must be carried out (personnel training);

(c) Performing the required routines day-in-and-day-out, faithfully and conscientiously, so that tableware and equipment are acceptable from a cleanliness standpoint and safe for use from the standpoint of public health.

2. Improper storage and/or any follow-up mishandling of cleaned and sanitized tableware and equipment items can, in an instant, destroy the desirable results outlined in paragraph 1. This would seem an unnecessary waste of management's money and dishwashing personnel's time and effort.

General rules for proper storage and handling include:

(a) A *clean* place for storage. Dusty or dirty shelves will quickly resoil cleaned and sanitized tableware or equipment.

(b) Storage areas must be dry and well above the floor. Splash from mopping solutions or food spillage, exposure to dust or floor dirt from sweeping must be avoided. Thus, pans or containers stored on low shelves should be inverted or covered. (A proper course to follow wherever these articles are stored.)

In many establishments, utensils and equipment items are hung in a self-draining position on conveniently located hooks or racks. Both hooks and racks should be constructed of a material with a surface that is resistant to the action of foods, cleaning compounds and sanitizing solutions.

Racks of cleaned and sanitized dishes/glasses, and utensils should not be stored, even temporarily, on the floor.

The food-contact surfaces of cleaned

Running water should be provided for the constant sanitation of ice cream scoops. Without the steady change provided by running water, the rinse will soon become contaminated.

and sanitized equipment must also be protected from splash, dust and other contamination. (This would include soiling by rodents or insects.)

(c) In addition to the above, suitable space and facilities should be provided for the storage of eating and drinking dishes and utensils to protect them from sneezing, coughing, or other careless contact by establishment personnel.

Flatware should be stored so that the handles are presented for personnel or customers to pick up for use. A correct employment of cylinders, as described in silverware washing, is a development which has helped protect flatware from recontamination by personnel.

The same care in removing cleaned and sanitized dishes and utensils from racks must be exercised by dishmachine personnel as was described in detail for the handling of these items by manual operators. Other foodservice personnel must be equally particular in the handling of sanitized tableware while preparing place settings, or whenever they use it.

Fingers should have no contact with any dish surface upon which food will be placed, nor with the lip-contact surface of any dish or utensil.

PERSONNEL

The establishment's food may be well prepared, delicious, and correctly served; but it will not be enjoyed, probably not even accepted, if served on unsightly dishware or accompanied by unclean eating utensils or spotted, filmed glassware.

In addition to the necessity for attractive tableware, the trained dish crew is well aware of the danger to public health should dishes and utensils be improperly sanitized.

The modern, trained dish crew member does, in fact, fall into the category of a skilled worker. He must know what to do as well as how and why he is doing it. In a machine operation he works with some of the most expensive equipment in the establishment. In any type of operation, he should be fast, skillful with his hands, sharp-eyed and alert to the many and varied requirements of his duties.

There should always be a supervisor, captain or "No. 1" man to whom the dishwashing employee can report any defects in his equipment and from whom he can obtain reliable advice in an emergency.

Modern dishwashing is not just "any job." It is a combination of skills. It is a team effort, with the team responsible to management for the proper performing of one of the most important activities of the entire foodservice operation.

"STERILE FILTH IS NO MORE DESIRABLE IN FOODS THAN UNSTERILE FILTH . . ."

. . .ANONYMOUS

SANITIZING QUIZ

1. True or False (check True or False):

	True	False
(a) A disinfectant destroys disease-causing organisms	( )	( )
(b) Sterilizing is the process used on dishes and utensils to destroy micro-organisms.	( )	( )
(c) Sanitizing and sterilizing mean the same thing.	( )	( )

2. Fill in the following statement:
pH plus a number shows whether the detergent solution is ________________ or ________________ in content.

3. Check correct statements.
   (a) "Sanitize" is the word used in foodservice to describe the killing of bacteria on dishes and utensils. ( )
   (b) A quick dip in 170°F. water will sanitize dishes and utensils in hand dishwashing. ( )
   (c) A bactericide destroys bacteria. ( )

4. Fill in the following statement:
In hand dishwashing, sanitizing is done in sink __________. Sanitizing may be done with ________________ at 170°F. or with approved ____________. If hot water is used, dishes and utensils must remain completely immersed for at least ____________. If approved chemicals are used, dishes and utensils must be completely immersed for____ minute. Two sanitizing chemicals used in hand dishwashing are ________ and __________.

5. True or False (check True or False):
   (a) In hot water sanitizing, dishes are placed in long handled baskets.
   True ( ) False ( )
   (b) After sanitizing, remove dishes immediately from baskets and stack them.
   True ( ) False ( )
   (c) Pots and pans and food-contact surfaces of other equipment must also be sanitized.
   True ( ) False ( )

6. Fill in the following statement:
A detergent-sanitizer is a detergent which has had a ______ agent added to its formulation. Two other names for detergent-sanitizer are ____________________ and ________________ ______________.

7. True or False (check True or False):
   (a) Chlorine, Iodine, Quaternary Ammonium compounds are 3 chemicals used in detergent sanitizers. True ( ) False ( )
   (b) When detergent sanitizers are used, the wash solution will not need changing so often. True ( ) False ( )
   (c) Detergent-sanitizer solutions are important in floor mopping.
   True ( ) False ( )

8. Fill in the following statement:
   (a) Dish and utensil storage areas must be ________ and well above the ________.
   (b) Cleaned and sanitized tableware and equipment should not be exposed to ________, ________, or ________ and __________ from sweeping.
   (c) Pans and containers stored on lowest shelves should be ________________

9. Check correct statements.
   (a) Careless handling and improper storage will recontaminate dishes, utensils or equipment. ( )
   (b) Racks of clean glasses may be stacked on the floor. ( )
   (c) Food-contact surfaces of cleaned and sanitized equipment must be protected from splash, dust and other contamination. ( )

10. True or False (check True or False):
    (a) Fingers should not contact any dish surface upon which food will be placed.
    True ( ) False ( )
    (b) Dishes can safely be stored on dusty shelves. True ( ) False ( )
    (c) Dishes and utensils should be air dried before storage. True ( ) False ( )
    (d) It is OK to "polish" glasses with a towel. True ( ) False ( )

Sanitized silver is offered with eating ends protected to be used with food protected by sneeze guards. View is from rear of service counter

12 Responsibilities for Dishroom Maintenance

The discussions of the different types of dishmachines and the various kinds of chemical compounds used in the machine's washing solution have given dishwashing crews a beginning background knowledge of the dishroom operation.

The next step is to learn what dishwashing personnel must do to help keep the operation working at peak efficiency. Just what are the responsibilities of the crew insofar as maintenance and cleaning are concerned?

Keeping the dishmachine operation at the performance level necessary for the production of sparkling clean tableware requires the alert and conscientious cooperation of dishwashing personnel and their supervisor. In many establishments, an engineer, handy man or other designated employee is available to take prompt care of minor repairs.

An important duty of a dishroom employee is the quick noting and prompt reporting of need for repair of the machine itself, or any problems involving his company's equipment or products.

In other areas, such as variations in water volume, pressures, temperatures, etc. the neglect of what may begin as a minor mechanical problem can easily result in an expensive, time-consuming shutdown of the machine while major repairs are made. Hence the good dishroom employee reports any change in the machine's operation, even if such a change seems only an unimportant inconvenience to him at the time.

A simple outline of the basic (main) activities of a dishmachine will help to point out the various areas of responsibility:

1. We begin with a wash tank containing a cleaning solution. The solution is made up of a correct amount of water for the model used, plus the properly measured amount of dishwashing compound.

2. In required volumes and with required pressure, the cleaning solution is directed, or aimed, at the tableware to be cleaned. The machine is constructed so that the cleaning solution strikes the ware according to a pattern which allows it to contact all surfaces of the soiled objects.

3. Next, hot rinse water is sprayed over the tableware in sufficient amounts and with enough force to remove any loosened soil or traces of wash solution clinging to the dishes. In many models, a final rinse sanitizes the dishware.

Responsibilities for Maintenance and Cleaning

In taking care of the responsibilities listed in item 1:

(a) Personnel are responsible for seeing that the wash tank is filled with the proper combination of water and dishwashing compound. The detergent dispensing system should be checked for good working condition. Before the first run of tableware, the wash tank thermometer is read, to be sure the wash solution is at the correct heat level.

(b) The dishwashing crew may regard the above steps as nothing but routine tasks. Nevertheless, as will be seen when item 2 is discussed, the exactness and care with which they are performed have a real bearing on maintenance and cleaning.

In taking care of the responsibilities noted in item 2:

(a) As already noted, variations in the required water volumes, pressures and temperatures are to be reported to management and corrected by a designated individual.

(b) Whether or not the wash solution strikes all parts of the dishware's surfaces is one of the chief concerns of the dishwashing crew.

(c) If the wash arms or sprays become clogged with mineral deposits, food particles, etc. the cleaning solution will not strike the tableware with sufficient force, or in the directions intended by the machine's manufacturer. In other words, the wash pattern has been changed and all surfaces of the soiled ware are not contacted by the cleaning solution. The result is a poor washing action and "dirty" dishes. Thin film and grease, etc. tend to cling to the dishes and the interior of the machine.

(d) Grease build-up can result in excessive foaming which can increase the level of deposits on the machine. Many modern detergents are formulated to control foaming and build-up of deposits in the machine. Personnel should still observe every sanitary procedure for keeping the machine's interior clean. Personnel help maintain cleanliness in the machine by careful scrapping and prerinsing of ware; by noting whether fresh water is being continuously added to the wash solution in the proper amounts, so that surface grease and scum overflow into the drain; and by changing water as necessary whenever it becomes too soiled.

The type and size of the machine will determine which procedures are to be followed to keep the wash solution acceptably clean.

(e) The mineral deposits referred to above occur when water is heated to high temperatures and the minerals carried in it are separated by the heat. This condition is called temporary hardness. The deposits collect on inner surfaces of the machine and in the wash arms and sprays, thereby reducing the effectiveness of the wash action.

(f) Dishwashing crews are apt to blame their detergent for most of these problems. And while it is true that varying conditions may require trying out different compounds before the most suitable one is found, this does not occur often enough to be considered a real problem.

(g) The discussion of dishwashing compounds (pp. 62-71) explains why the various chemicals were included in the products to control water hardness, film, grease, foaming, etc., to condition the food soil for quick removal and holding up in the wash solution until its flushing off by the rinse.

The well balanced dishwashing compound can do much to maintain cleanliness in the machine, always provided it is used in the recommended concentrations and with the washing solution at the required temperature, pressure and volume. Hence the importance of the preliminary steps listed under item 1, on the preceding page. During these procedures, personnel can note and report any defects in the beginning of the operation. Any such defects could well become a maintenance problem.

(h) The mineral deposits caused by temporary water hardness (referred to as "scale") are not always entirely controlled by the detergent. Clean-up procedures, to be described later, must be carried out by personnel.

To carry out the responsibilities listed in item 3 on the preceding page, these steps are required:

(a) The rinse jets, too, may become partially, or sometimes entirely, clogged by mineral deposits, in which case the rinse water will be discharged at an angle which provides either partial or no coverage of the tableware. Power rinse jets should be checked by personnel to make sure they are free of food wastes or particles. Final rinse jets, be-

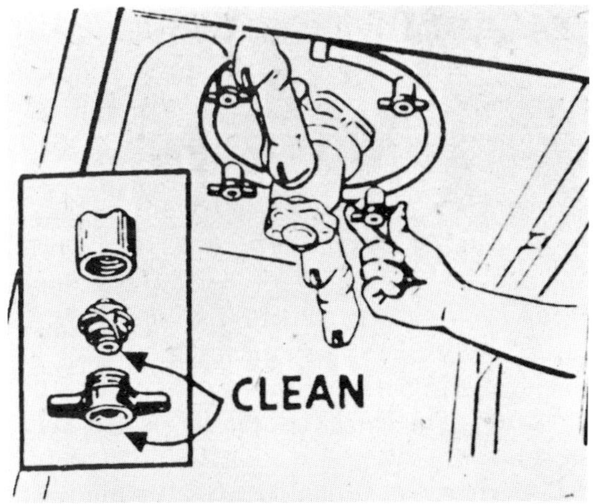

Rinse openings clog from sediment in the hot water system. . . and a clogged sprayer means that some of the dishes will not be properly rinsed. Thorough rinsing is important and possible only when all of the rinse sprayers are open.

ing smaller, may often become clogged by hard water scale deposits. (A descaling type rinse additive, capable of keeping the final rinse compartment free of dried-on hard water salts has been used to advantage in hard water areas. Refer to the section on Dishwashing Compounds.)

(b) Water spots and film on dishes can result from inadequate rinsing. A too short rinse time, a too high rinse water temperature, or too low or high water pressures are all conditions requiring the attention of the proper maintenance personnel.

Importance of Maintenance and Cleaning

Many dishwashing personnel believe their chief duty to be the actual cleaning of dishes, utensils, etc. They consider other tasks to be of secondary importance. However, when employees understand the outline of normal dishmachine activity, they can more readily appreciate the necessity for devoting as strict and conscientious attention to machine cleanliness and maintenance as they do to ware washing procedures.

The simple facts are that: (1) a dirty machine will usually produce dirty dishes; (2) a dirty machine can lead to major mechanical problems, a shutdown of the machine for repairs, and, should such repairs be made during operating hours, a probable inconvenient and difficult work situation for personnel.

CLEANING THE DISHMACHINE

1. Official health agency regulations plus continuous efforts of dishmachine manufacturers to provide industry with equipment that is modern, long lasting, offers better performance and minimum maintenance have contributed much to the ease and efficiency with which personnel can accomplish cleaning routines. Note the emphasis placed on "easy cleaning" in the following examples:

(a) The 1962 U.S. Public Health Service Food Service Sanitation Ordinance under Methods and Facilities for Washing and Sanitizing states: "Dishwashing machines shall be of such materials and so designed and constructed as to be easily cleanable."

Also, where floor-mounted equipment is located next to a wall but not sealed to it: "sufficient space shall be provided to facilitate easy cleaning between, behind and beside all such equipment."

State and local food establishment regulations contain similar requirements.

(b) The National Sanitation Foundation's Standard No. 3, relating to commercial spray type dishwashing machines is also specific about the cleanability factor in machine construction and design. (The individuals who prepared this Standard included representatives of official agencies, public health consultants and leading equipment manufacturers.) Note the following requirements from Standard No. 3:

(i) "It is required that the bottom of the machine's main structure shall be high enough above floor level to permit cleaning beneath and in no instance shall be less than 6 in. above the floor."

(Exception allowed for necessary piping connections and gas burner enclosures. Such enclosures should be at least 4 in. above the floor.) This is in accord with USPHS recommendations.

(ii) Summarizing further legs and feet of machine must be so shaped at floor contact as to prevent the accumulation of dirt and harborage of vermin. Legs and feet should be of simple design with no uncleanable areas.

(iii) "All splash contact surfaces of the prewash, wash, pumped rinse, and final rinse areas shall be readily accessible and easily cleanable, either in an assembled position or when removed. Demountable parts shall be readily removable. When the design of the dishwashing machine is such that on installation such surfaces are not readily accessible and easily cleanable, separate access openings shall be provided to facilitate easy cleaning."

"Readily accessible" is defined as being easily exposed without the use of tools for proper and thorough cleaning and visual inspection.

"Easily cleanable" means readily accessible and of such material, finish and so fabricated that soil may be effectively removed by normal cleaning methods.

These are only a few examples of the emphasis health officials and dishmachine manufacturers place on the necessity for thorough machine cleaning. To make sure equipment is kept clean, they make available machines so designed and constructed that the cleaning routines are as simple and easy as possible. It only remains, then, for dishwashing personnel to carry out the required cleaning with exactness. There is no place in the modern dishroom for careless, slipshod cleaning procedures.

CLEANING THE MACHINE AND DISHROOM

The dishcrew's supervisor should be made responsible for the machine's cleanliness, even though he may take no active part in the cleaning routines. In small establishments where no supervisor is employed, the manager should inspect the machine thoroughly, inside and out, several times a day and always at the end of the day's operation.

Cleaning procedures for the machine and surrounding areas are carried out on a daily and weekly basis.

Daily Cleaning Steps

1. Turn off the detergent dispenser and rinse injector, if these are used.

2. Turn off machine's heat sources.

3. Open the drain valves on all tanks and remove soiled water. If this is done while the water is still hot, less surface scum will adhere to the walls of the wash tank.

4. Remove wash arms. Flush out interior of the machine with hose and hot water to remove any food debris. When all demountable parts are out of the machine (the next steps), it is suggested that a brush be used to scrub and clean the inner parts of the machine. Detergent spray units are also convenient to use. This will remove any adhering materials, prevent odors, reduce rusting and promote interior drying when the machine is not in use.

The overflow pipes from the various tanks should be clean and free of debris. Otherwise, the clogged overflow allows the water level to rise to a point which interferes with the proper action of the lower wash arms and jets.

5. Flush out wash arms to remove any food scraps or debris (toothpicks, cigarette butts, paper, etc.) which may be clogging them. A shower head scrapper is good for this purpose. Double check to be sure both wash arms and nozzles are completely clean after they have been flushed out.

Under heavy load conditions, it will probably be necessary to flush out the wash arms after each meal or even more often, especially in establishments where the machine is operated almost continuously. In the modern dishmachine, a worker can remove and replace the wash arms with little effort in a very few moments. No tools are required.

The arms in the power rinse section should also be removed and flushed out.

6. Remove the scrap trays, flush and scrub clean. Make sure all openings are free of any refuse. Trays should be in good repair so that large food scraps cannot enter the wash tank. We can understand now why thorough scrapping and pre-rinsing are so important and time-saving.

Scrap trays may need cleaning during a dish run; but they should always be taken care of after each meal as well as at the end of the day.

7. Remove curtains. These may be cleaned by laying them on a dish table, wet, sprinkling with a detergent, scrubbing with a good brush and rinsing. This should be an end-of-the-day procedure. Curtains washed daily are free from odor. They will last longer, since dissolved food materials which might stain or rot them are regularly removed.

The curtains should be left off overnight and the doors of the machine left open to promote drying and allow circulation of air through the machine.

In those establishments where the dishmachine is in use almost 24 hours a day, The National Sanitation Foundation recommends that two sets of curtains be provided for use on alternate days. This allows one set to dry out thoroughly while the other set is being used.

8. Machine manufacturers suggest that rinse jets be cleaned in place unless one employee can be made responsible for their correct placement. In other words, one member of the dish crew—the No. 1 man—would be trained by the manufacturer's field representatives.

If the rinse jets are not put back in exactly the same position as when installed, the spray pattern can be changed enough to

Sweep floors only during slack or closed periods using a good broom.

be ineffective.

Rinse jets should be checked daily to be sure they are not materially reduced in size due to mineral build-up, or plugged with scale particles from the hot water supply line. This condition can be corrected by deliming. The deliming process is discussed under periodic procedures.

9. The exterior of the machine can be kept spotless, shiny and new looking by cleaning with a mild detergent solution, rinsing with clean water and damp drying with a clean cloth. Do not clean while the surface is hot.

Stainless steel may become discolored from food spillage. In this case, detergent manufacturers can provide mild acidic type products to remove the discolorations. These products, even though mildly acidic, should be used with caution. The manufacturer's use instructions should be followed exactly. When in doubt, the detergent representative will advise and demonstrate proper usage.

10. The detergent dispenser and rinse injector should not be neglected when cleaning. This equipment is usually on loan from the detergent manufacturer, hence his representative's cleaning instructions should be followed in order to prevent any damage or adjustment change while cleaning.

11. Soiled and Clean Dish Tables, Scrapping Sinks, etc.—Food wastes should be wiped up regularly during time of operation. Thorough cleaning after meals and at the end of the day should be accomplished by the same methods used for exterior of the machine.

12. Floor—The dishroom floor should be constructed of a smooth, easily cleanable material which resists the seepage of water and grease, etc. into its surface. The floor should be sloped sufficiently to provide water run-off to drain.

This floor must be well maintained by personnel. During peak operating periods, a quick sweeping or mopping may be all that time permits. After each meal, however, and at the end of the day, a thorough cleaning is necessary. Any food residues left after a casual rinsing will result in dull floors and slippery areas.

It is recommended that scrubbing be done using a germicidal detergent solution (a combination of detergent and disinfectant). In this way, the value of sanitizing is gained with no loss of cleaning efficiency. Standard cleaning procedures for the particular type of floor should be followed.

Personnel should be instructed in efficient cleaning routines; cleaning and rinsing a small floor area at a time; proper use of cleaning tools, etc.

Weekly Cleaning

Garbage cans—Scrub out thoroughly. Do not neglect the lids. Use an odor formation preventive product periodically, as instructed by product representative.

Give all the equipment a thorough overall cleaning to cover legs of machine, work tables, sinks, backs of equipment, clean storage shelves and carts and any other areas overlooked during the daily clean-up. Light fixtures should also be checked and cleaned.

In many establishments, complete wall washing is included in weekly procedures, rather than on a more infrequent basis.

Clean dishroom walls and ceiling (a) contribute much to the sanitary atmosphere of the area, (b) keep the room lighter and (c) promote efficient personnel work habits.

Periodic Cleaning

Deliming—Lime deposits must be removed regularly from the machine.

Their presence makes proper cleaning difficult, prevents thorough rinsing and thus reduces the machine's efficiency. Mineral deposits also cause rough surfaces to form inside the machine and these provide breeding places for bacteria.

Responsible manufacturers can provide safe acid cleaners which can be recirculated to keep the machine free of lime. A spray cleaning unit can also be used with an acidic detergent concentration. The acidic detergent removes lime scale and film. The spray makes it easier to reach areas not wholly accessible to the employee who cleans the machine.

The dishroom supervisor or some other staff member should personally direct the deliming procedure to make certain that it is done properly and on schedule.

Walls and ceilings not cleaned weekly should be washed before discoloration or dullness detracts from the desired atmosphere of over-all dishroom cleanliness.

LIGHTING AND VENTILATION

The dishwashing area must be well lighted and ventilated if maximum efficiency is to be expected from the dishwashing personnel. Good visual inspection of tableware is impossible with poor lighting.

Steam and heat in the dishroom, caused by the constant operation of the dishmachine, must be removed by mechanical exhaust ventilation directly over the machine, or by general room ventilation. Mechanical ventilation tends to reduce the humidity in the room and thus aids the air drying of the dishes. Too much ventilation may affect the wash, power rinse and final rinse temperatures and should be avoided. The overall dishroom ventilation should provide 60 changes of air per hour.

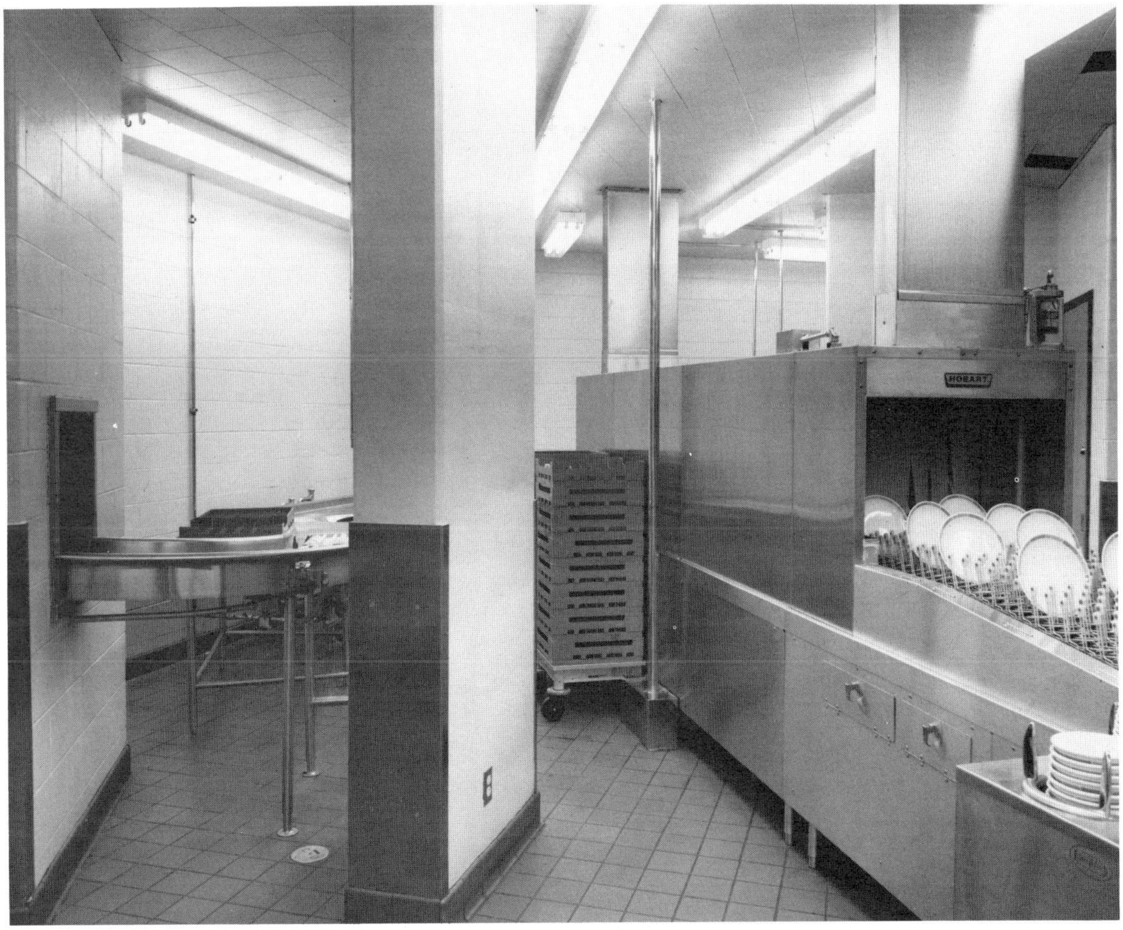

Plenty of light makes it easy to inspect clean dishes before they go into self leveling dish dispensers. Soiled dishes come into well lighted area by conveyor.

RESPONSIBILITIES FOR DISHROOM MAINTENANCE QUIZ

1. Fill in the following statement:
   If the wash arms or sprays become clogged with mineral deposits, food particles, etc., the _______________ will not strike the tableware with _______________ or in the _______________ intended by the machine's manufacturer.

2. True or False (check True or False):
   (a) "Temporary hardness" occurs when water is heated to high temperatures and the minerals carried in it are separated by this heat. True ( ) False ( )
   (b) Clogged rinse jets cannot do a satisfactory rinsing job.
   True ( ) False ( )
   (c) Dishwashing personnel should be interested only in the cleaning of dishware.
   True ( ) False ( )
   (d) A dirty machine usually produces dirty dishes. True ( ) False ( )
   (e) Dirty machines do not create maintenance problems. True ( ) False ( )

3. Check correct statement.
   (a) Modern dishmachines are designed to be easily cleanable. ( )
   (b) If thoroughly flushed out with hot water, it is not then necessary to scrub the interior parts of the machine. ( )
   (c) Under heavy load conditions, the wash arms will probably need to be removed and and flushed out after each meal, or oftener. ( )
   (d) After scrubbing curtains, put them back on the machine. ( )

4. Fill in the following statements:
   (a) Rinse jets should be checked _______________ to be sure they are not plugged with scale particles from the _______ _______ _______ _______.
   (b) Stainless steel may become _______________, due to food spillage. Mild, acidic type products will remove the _______________.

5. True or False (check True or False):
   (a) Cleaning personnel should not touch detergent dispenser or rinse injector. These are the property of detergent manufacturer. True ( ) False ( )
   (b) Soiled and clean dish tables, scrapping sinks, etc. should be wiped off regularly during operating period, cleaned thoroughly after meals and at end of day.
   True ( ) False ( )
   (c) A germicidal detergent should be used in cleaning dishroom floor area.
   True ( ) False ( )
   (d) Garbage cans and lids need thorough scrubbing and hosing, inside and out, at least once a week. True ( ) False ( )

6. Check correct statements.
   (a) Cleaning light fixtures is not the responsibility of the dishwashing crew. ( )
   (b) Clean dishroom walls and ceilings keep the area lighter, make it easier to work efficiently. ( )
   (c) Lime deposits must be removed from the dishmachine regularly. ( )
   (d) Deliming the machine is done by the detergent company's representative. ( )

7. Fill in the following statement:
   The dishwashing area must be _______ _______ and _______ _______ if the dish crew is to work at peak efficiency.

8. True or False (check True or False):
   (a) Alert dishwashing personnel should note and report any changes (even minor) in the machine's operation. (True) False ( )
   (b) Changes in water temperatures, volume or pressure are very important, could lead to major repairs. True ( ) False ( )
9. Check correct statements.
   (a) Care in the performance of routine procedures will make maintenance and cleaning easier. ( )
   (b) Once the detergent dispenser has been filled, it needs no additional checking. ( )
   (c) A well balanced dishwashing compound can help maintain machine cleanliness, provided it is used in correct concentrations, and at correct temperature, pressure and volume. ( )
10. True or False (check True or False):
   (a) Personal cleanliness, the cleanliness of the dishmachine and work area are the signs of a well trained, efficient dishwashing crew. True ( ) False ( )
   (b) The dishmachine is the most important "tool" used by the dishwashing crew. If clean, sanitized dishes and utensils are to be regularly produced, the dishwashing crew has to devote time and care each day to the cleaning of their dishmachine.
   True ( ) False ( )

The personnel responsible for keeping this dishroom clean would earn high praise in any foodservice operation.

Notes to Instructors:

The Personal Hygiene class session is the logical place to finish the demonstration-experiment introduced near the conclusion of the Bacteriology chapter. (Review description on page 14.)

The covered Petri dishes containing nutrient agar (bacteria food), and bacteria placed thereon by coughs, hands, or swabbing of selected objects, have been left in a warm place at temperatures suitable for bacterial growth, for several days. By this time, colonies of bacteria growing on the agar should be clearly visible to the naked eye.

This visible evidence of the existence of life forms normally invisible to the naked eye may be presented effectively at any one of several points during the Personal Hygiene lesson. Suggested are the hands habits discussion, the listing of correct personal hygiene habits, the suggestions for cooperation with management and personnel's responsibilities as foodservice workers.

In preceding lessons the class has learned that some of the invisible life forms are disease-causing. They have been told that they (the foodservice workers) can pass certain of these diseases on to fellow-workers and/or customers by failure to carry out the easy-to-understand sanitation rules designed to protect both human beings and food.

It is very important for personnel to appreciate the meaning of this demonstration-experiment. Its purpose has been to prove the existence of living bacteria on our persons and on objects all around us. This has been accomplished by exposing bacteria to suitable food and temperatures favorable for rapid growth and multiplication of the organisms.

True, these conditions do not exist in every work-a-day situation involving foodservice employees. But having learned about the foods and temperatures bacteria prefer for growth and multiplication (See Food-Borne Diseases, Food Protection), personnel can now visualize what can—and does—occur far too often in food to which disease-causing bacteria have gained access. How then, can some foodservice workers neglect to practice the simple food protection procedures? How can careless hands habits be tolerated by fellow-workers or customers?

13 Personal Hygiene

THE PERSON IS THE THING

This chapter presents a detailed discussion of the personal habits of the foodservice worker, including some considered undesirable and others which should be cultivated.

In addition, the importance of the individual foodservice worker in relation to his responsibility in the food sanitation program will be explored. Hence the subtitle of the chapter—*The Person Is the Thing.*

Why is the sanitation trained foodservice worker important?

1. He knows what bacteria are, how they live, grow and multiply.
2. He has learned how some types of illnesses are communicated when certain disease-causing bacteria are transferred directly or indirectly from sick people to well people.
3. He has learned why and how food must be protected in order to prevent food intoxications and food infections.
4. He knows the importance of proper cleaning methods and the necessity for correct sanitizing procedures.
5. He understands the part played by insects and rodents in disease transfer and appreciates the need for pest elimination and continuous control measures.
6. Most important of all, he realizes now that bacteria are living organisms and that he is inescapably involved in coping with these organisms during his on-the-job working hours.

Adequate refrigeration, proper cooking routines, correct washing and sanitizing of tableware and equipment, control of insects and rodents, all of these things could be done and it would still be possible to have disease transference in foodservice operations if the foodservice worker were to forget or neglect to follow the simple and common sense rules of sanitation he has learned. These rules must be remembered and acted upon every day.

This is why the person is the thing that counts in a food sanitation program. The success or failure of the program actually depends upon the attitude and cooperation of the foodservice personnel involved.

It follows that the foodservice worker must be competent. The competent employee makes a definite contribution to his place of work.

The incompetent employee can be responsible for any number of insanitary conditions in the establishment—conditions which may lead to the illness of some customers or fellow workers. Such incompetence can eventually mar the favorable image of any establishment, and substantially lower the customer count.

Competency involves two things:

First, the employee must know his job. He must be aware of his duties and must perform them so efficiently that the foodservice operation will progress smoothly.

Second, his efficiency must include the carrying out of sanitation rules as they pertain both to the job and to the employee.

Foodservice is exacting work, impossible to perform efficiently unless the worker is in good physical condition. In addition, the sanitation-trained worker knows that the state of his health may have a direct influence on the health of fellow workers or those who avail themselves of the operation's facilities.

Therefore a few basic rules should be followed to keep the employee physically fit for work. They are as follows:

The well-groomed waitress.

1. Have a thorough physical check-up once a year. This check-up will not guarantee a day-by-day freedom from infections which may cause food-borne diseases. But such an examination will provide the employee with necessary information on his general physical condition.

2. Visit a dentist a couple of times a year. Infections of the teeth and gums could lead to serious trouble elsewhere in the system. And no one can do an efficient job with a nagging toothache.

It is also suggested that periodic eye examinations be included in the physical check-up program. Headaches from eye strain, neglected corrections in glasses, or other eye problems perhaps not realized by the individual prevent a foodservice worker from carrying on the daily routines with efficiency.

3. Eat proper amounts of the right foods.

4. Get plenty of rest at night.
5. Stay at home when feeling sick.

While these general rules could apply to anyone in any line of work, the foodservice employee, above all others, needs to observe those requirements of personal hygiene and cleanliness which result in a well-groomed, well-poised worker. Requirements include:

1. A daily bath. This is a "must."

2. So are frequent shampoos. Dirty hair in a food establishment is as offensive as an unclean body.

3. Teeth should be cleaned daily. Many foodservice workers keep a toothbrush in their lockers for use on the job.

4. Use a deodorant and a mouthwash. Then the foodservice worker can be sure fellow employees or customers will not be offended.

5. Underclothing must be fresh and clean daily. Perspiration and food odors penetrate underclothes and can be instantly detected.

6. Clean uniforms are a requirement. Keeping them unsoiled takes effort and experience in "working clean" by learning not to wipe the hands on the uniform.

7. Wear sensible, comfortable shoes at work. They should be kept clean, polished and in good repair. It is impossible to do an efficient job in unsuitable shoes—and unsuitable means too high heels, rundown heels, worn-out soles which make slipping and falling easy, or sloppy shoes of the bedroom slipper variety. This applies equally to kitchen and dishroom employees.

8. Female workers are required to wear hair nets. Male chefs, kitchen workers, food runners, dishwashing personnel, etc. wear caps. Falling hair has no place in food.

9. Avoid wearing jewelry on the job. It does nothing for the appearance of a pro-

Personal hygiene stresses importance of daily bath.

fessional foodservice worker. In addition, pieces of jewelry can break, fall into food and find their way to a customer's plate. Damage claims have originated from just such occurrences.

10. Hose should be worn, should be well fitting, with no sagging or visible runs. Gum chewing by food servers or counter attendants can spoil the picture of efficiency. It should not be indulged in while on duty.

11. Hands and nails should be clean. Nail polish should be colorless or natural in shade. The nails and cuticle should receive frequent care to avoid hangnails.

12. Face make-up should be applied in moderation.

The rules and suggestions outlined thus far have covered physical fitness and the daily physical preparations for the job. In addition to these, management expects courteous, pleasant service to be given to customers. Under some circumstances this can be difficult. Nevertheless the waiter and waitress or counter attendant are the persons who offer the hospitality of the house to the customer. They are the substitute for the hostess who dispenses hospitality in the home. As such, they can either bring the customer back or drive him away for good. From a practical standpoint, the well-groomed, courteous, efficient worker with the pleasant personality is the one who can, most often, increase his or her income materially with numerous tips.

Finally, there are the "special" requirements for the foodsergice worker, the requirements which must be observed in order to break any possible chain of infection in the operation. These requirements apply to anyone in the establishment whose work involves contact with food, utensils, or equipment.

The carrying out of these requirements demands constant thought and alertness. Some of the routines may involve breaking personal habits of long standing. The breaking of an established habit is a difficult process for most adults at any time.

Hands habits have been mentioned several times in previous chapters. A summary of the correct hands habits the sanitation trained foodservice worker must learn emphasizes the following practices:

(1) Wash hands thoroughly after restroom visits, after smoking or after handling any unclean object before again working with food or food utensils. This practice should become automatic.

(2) Keep the hands away from the mouth, nose, ears, eyes and hair while on the job. If any of these areas are absentmindedly handled, wash the hands before resuming duties.

Don't cough or sneeze on food or dishes.

3. Cover coughs and sneezes—and wash hands if spray from the coughs or sneezes has contaminated them. If plastic gloves are being worn when coughs or sneezes are covered with the hands, discard the gloves for a fresh pair.

4. Immediately clean, disinfect and

bandage any cuts or sores on the fingers, hands or arms. Bacteria from soiled dishes, food remnants, etc. may infect these cuts. In addition, bacteria from unbandaged and infected cuts and sores may be transferred to food and cause illness to fellow workers or customers. Hangnails are included in this category.

The wearing of plastic gloves provides an additional safeguard against contamination of fresh food, clean dishes, utensils, etc. by infective material which might seep through a bandage. Should the gloves be punctured—as from the tines of a fork—protection no longer exists. Bacterial infection could pass through the puncture hole to contaminate food, utensils or equipment. Punctured gloves should be discarded for a new pair.

5. After removing soiled dishes from tables, after wiping up dirty tables or after cleaning restrooms, the hands must be washed. Soiled objects have been touched which have been in contact with the mouths of other people or contaminated in various other ways. Such contamination must not be transferred to the foodservice worker by placing the fingers in the mouth or touching other facial areas.

6. While working at the preparation of raw salads or any food to be eaten without cooking, any bacteria on the hands, under the nails, etc. can be transferred easily to the food. Therefore, it is absolutely essential that the hands be clean before such food preparation is begun and that no contamination of the hands occurs during the working period. If it should, the hands must be washed before resuming work.

7. The fingers must not be dipped into food in order to taste it. Use a spoon, and never put the spoon back in the food after it has touched the mouth. Another habit too often noticed is that of dipping the hands into ice bins where ice intended for cold drinks is stored.

One more reminder about hand washing. The thorough washing practiced by foodservice workers does not mean a hurried trip to the wash basin, a quick rinsing of the finger tips under cool water and a hasty shaking dry. Hand washing for foodservice personnel *does* mean a *careful, thorough* washing.

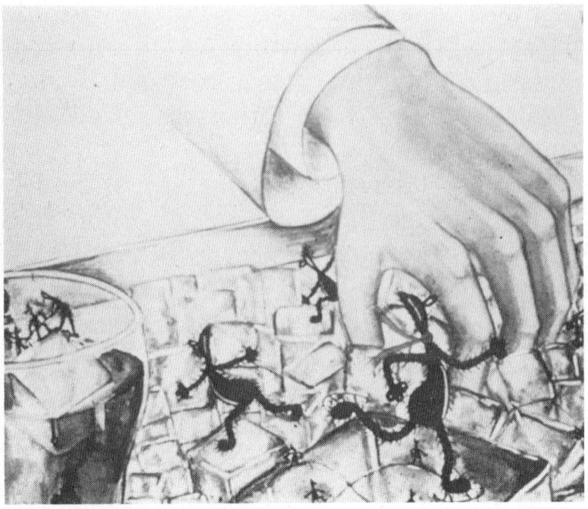

Keep fingers out of ice; use tongs or scoops. When ice intended for customers' drinks is picked up by the fingers, soil and bacteria present on the hands can be transferred to the customer's drink.

When lipstick stain remains on cup after washing, worker unloading dishmachine should reject cup.

Egg remains on fork after washing, a reminder that both dishwashing and service personnel should check dishes and utensils carefully before service.

Cooperation with Management to Raise Sanitation Standards

1. Examination of eating and drinking utensils before service. This is the task of both dishwashing and serving personnel. Cups with lipstick visible on the rims, glasses or utensils with water spots, utensils soiled

with food residue have a most unpleasant effect on customers who are influenced by what they see, not what they cannot see.

The personal responsibility of the food-service worker is to avoid serving the customer with what he cannot see, but which may be much more dangerous to him than obviously soiled utensils. Do not recontaminate dishes or utensils after they have been properly cleaned and sanitized.

Right way to handle cups.

Wrong and right way to serve cups and saucers. Do not contaminate drinking surface by stacking cups as at left. Handle saucers as at right.

2. When management provides garbage cans with tight fitting lids, keep the lids on the cans, replacing them each time after use. Keep refrigerator doors closed. Use storage facilities properly.

3. Obey the "No Smoking While on the Job" rule. Fingers come in contact with the lips while smoking and the hands are contaminated by any germs present in the mouth. These germs can be transferred to food by hands which have been in contact with a cigarette. The "no smoking" extends to storeroom personnel for some of the food handled there is not intended for cooking before service.

Cooperate with management to maintain high standards of sanitation. Keep food protected.

Keep all perishables in the refrigerator.

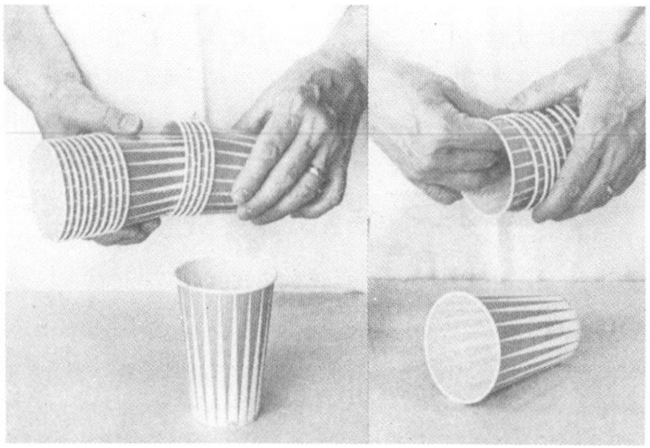

Correct and incorrect method of handling paper cups. Never grasp a paper cup by the end which will contact someone's mouth as at right. Handle paper cups as at left. Keep fingers out of inside of clean cup at all times.

The foodservice worker actually carries a tremendous responsibility because he can, in the course of his daily duties, either spread disease or prevent the spread of disease in his place of work.

It is said that poor foodservice practices can generally be laid to three causes:

1. Lack of knowledge.
2. Carelessness.
3. Unwillingness to cooperate and perform job routines in a sanitary manner.

After a study of this text, the foodservice worker has the necessary knowledge.

The careless worker is simply not thinking. Failure to improve, or persistence in, incorrect hands habits indicates a general lack of interest in the performance of any routine job duty. Employees with this attitude are holding themselves back from any advancement, and are often discharged.

The employee who has knowledge, is not careless, but is unwilling to cooperate with management in a sanitation program should ask himself if he belongs in the foodservice industry. The modern foodservice worker must be sanitation conscious in order to satisfy the requirements of his profession as well as to safeguard his own and other persons' good health.

Again, *the person is the thing.* Management cannot state that employees are trained simply because they have sat through a training course. The proof comes in the improvement noted in sanitary foodservice practices, the development of correct hands habits, the interest in doing things in the required way. Management and the health department can assist and advise, but the success of the program rests on the shoulders of the foodservice workers.

In 399 B. C. Socrates said: "One cannot get nearer to God than by giving health to his fellowmen."

This would be a good motto for foodservice workers to remember as they go about their daily tasks.

Note to Instructors:

One of the most effective demonstrations ever used to show foodservice workers that invisible bacteria do exist and can be transferred from place to place involves the employment of a black light, a fluorescent paste and a jar of vaseline to mix with the paste.

A black light is used for filtered long-wave ultra-violet radiations. Such lights are frequently used by health departments in pest control programs to detect the presence of rodent wastes. It is possible the light could be borrowed from a local or state health department. In some areas they are available for rental or purchase.

A short skit is performed. Stage props include a table, china, glassware, silverware and a towel.

The instructor must apply the vaseline-fluorescent paste mixture to his finger-tips just prior to the demonstration, preferably without the class being aware of what he is doing. This could be accomplished by stepping out of the room or moving over to one side while a student assistant sets up the demonstration material.

The instructor then launches into an all too common situation experienced by foodservice personnel. He can inject enough humor to make his points emphatic without causing any resentment.

He becomes, temporarily, a waitress whom we will call Mary.

Mary oversleeps, dresses hurriedly, arrives at work with barely enough time to don her uniform, visit the restroom, put on her hairnet and refresh her make-up. To increase her problems she is catching a cold and has forgotten her supply of tissues.

An amusing description and appropriate gestures will bring sympathetic chuckles from the class. Too many of them have been in Mary's predicament and they are in sympathy with her even though they are laughing.

Because she is late, Mary's preparations in the locker room have been hurried and she leaves without washing her hands. She arrives at her station and proceeds to set up her tables under the reproving glare of a supervisor. Because she is hurrying she becomes nervous and carelessness is the result.

The instructor playing the role of Mary will go through the motions of setting the table. Since the sketch must be brief, his actions will be exaggerated. But he will grasp the plate with his thumb well inside the food-contact area; his fingers will be on the drinking rim of the glass; he will pick up the cup by the rim rather than by the handle; his hands will also contact the ends of the silverware which will enter the customer's mouth. From time to time he will scratch his head, put a finger in his mouth, rub his nose. Finally, he will flick the side towel across his forehead then deftly polish the area around the place setting, remove an imaginary speck of dust from the plate.

While doing all these things the instructor keeps up a running patter of talk with an imaginary waitress friend at an adjacent station—last night's date, oversleeping, the sniffly nose, etc.

When he has finished, the instructor surveys the results of his labors. At this point the class expects to be asked to name the mistakes made in the handling of the dishes and utensils. Instead, the instructor asks how the set-up looks, if it is ready for the first customer. The class agrees that it is.

What about the incorrect things he did? Will the customer know? No, of course not. The table, china, glass and silver appear gleaming and shining clean.

Now the instructor reviews all that Mary did before appearing on the floor. He mentions her sniffles and her omission of handwashing. Has Mary contaminated this set-up by her carelessness?

True, bacteria are invisible. But there should be some way of learning whether or not Mary's hands have left traces—contamination—which might conceivably spread the germs of disease.

Yes, there is a way. The classroom is darkened, the black light turned on and passed slowly over the set-up. Immediately, every place where Mary's fingers have been reflects a brilliant white color. Thumb prints on the plate, finger prints on the glass, cup and silverware, smudges on the table and towel, marks on the instructor's forehead, hair, nose and mouth.

The effect of this on a class totally unprepared for such a climax is invariably quite startling. When the lights are turned on, the instructor faces a group totally receptive to the fact that simply because something is invisible, there is no use in pretending it does not exist or disregarding its dangers.

The demonstration actually takes less time to perform than it does to describe it. One caution—during the demonstration, use a linen or dish towel which has never been commercially laundered. Professional laundries use fluorescent laundry marks which will become visible under the black light and mar the effectiveness of the demonstration.

PERSONAL HYGIENE QUIZ

1. Give 3 basic rules for physical fitness: ______________________
____
____

2. Give 3 basic rules of personal hygiene ______________________
____
____

3. Fill in the following statement:
The foodservice worker must be very particular about his ______________

4. Check correct statements.
(a) Don't try to hold back a cough or sneeze. It isn't good for you. ( )
(b) Cover cough or sneeze with *hand.* ( )
(c) Use disposable tissue. ( )
(d) Use handy towel or apron. ( )
(e) Wash hands when convenient. ( )
(f) Wash hands before working again with food or cleaned utensils ( ).

5. A customer has been given a dirty fork. He angrily calls it to the attention of the waitress. She should: (Check correct statements)
(a) Explain it isn't her fault, she doesn't wash silver. ( )
(b) Tell him about the terrible dishwashing help. ( )
(c) Report incident to supervisor. ( )
(d) Apologize, bring a clean fork. ( )
(e) Later, tell the dishwashers what she thinks of them. ( )
(f) Next time, check silverware and dishes before serving them. ( )

6. As a kitchen worker, you want a raise. You tell your new supervisor:
(Check correct statements)
(a) You have worked one year, long before he arrived. ( )
(b) You are faithful, are on the job no matter how ill you feel. ( )
(c) You never stop, sometimes even eat on the run while working. ( )
(d) Because you are well trained, follow new rules, understand required routines, you feel able to take on more responsibility, with higher pay. ( )

7. As a cafeteria salad maker, you have prepared, by hand, the tossed salad for lunch. Now you realize a bandage is missing from a cut finger . You should:
(Check correct statements)
(a) Inform your supervisor. ( )
(b) Using proper utensils examine *carefully* any tossed salad remaining in kitchen. ( )
(c) Alert the salad servers on the line. ( )
(d) Say nothing. Perhaps no one will find it. ( )

8. What are the 3 main reasons for poor foodservice practices by personnel?
____
____
____

9. Here are more suggestions involving personal hygiene habits, cleanliness and appearance. Do not repeat answer given for question 2.
   Fill in the following:
   (a) ______________________________ must be fresh and clean daily.
   (b) Wear ______________, comfortable ______________ at work.
   (c) Also wear a ______________ or ______________ while on the job.
   (d) Do not wear ________________ while on the job.
   (e) Apply ____________________ in moderation.
   (f) Keep ______________ clean.
   (g) Keep ______________ and ____________ clean, well cared for.
   Use ________ ________ that is colorless or natural in shade.
   (h) Wear ______________ with no visible ______ and straight ________.
   (i) Do not __________ __________ while on duty.
   (j) No ____________ on the job.

10. Do you consider your personal habits important to your job? Yes ( ) No ( )
    If so, give three reasons why. ______________________________
    ____
    ____

Hairnet, clean apron and uniform, clean hands in clean plastic gloves offer protection to food items as they are prepared. The person is the control in the chain of tasks that leads to service of safe food.

14 Outlines for Follow-Up Meetings

FOLLOW-UP MEETING NO. 1

General

The over all purpose of follow-up sanitation meetings is to keep alive and further stimulate personnel interest in the sanitation program. This entails a frequent review of material already covered plus the introduction of new material to enlarge the scope of personnel knowledge.

Management-supervisory staff meetings are a routine procedure in many establishments. But the fact that personnel, too, can learn much from a series of informal sessions on operational methods is not always realized. How *much* is learned will depend on the content of material and the method of presentation.

To maintain interest, review material must be presented from continually fresh viewpoints with varied visual aids. New and more in-depth discussions of subject material previously covered in class sessions, reference to related topics—all these will both increase personnel knowledge of food sanitation, create a realization of the subject's importance and inspire personnel to participate actively in the program by correcting faulty habits and substituting correct procedures.

As soon as possible after conclusion of the initial training course, management should begin to collect various suitable visual aids for use in follow-up meetings. These could include moving pictures, film strips, posters, pamphlets, charts, booklets, leaflets, pass-out sheets, etc. These materials are usually available for purchase or rental (films); in some cases free samples of certain printed matter are provided.

The local health department will be very cooperative with suggestions and loan of any films they have on hand. They may be able to help with a projector and screen. On occasion, a health department employee has operated the projector or taught an establishment supervisor how to do this.

Contact your State Health Department to learn what films are available in its Film Library. The library should also be able to provide listings of moving pictures and other materials offered by the various governmental Health Agencies.

The Superintendent of Documents, U.S. Government Printing Office, Washington, D. C. 20402 will provide complete film listings, and it would be advisable to get on this department's mailing list. Allow ample time for a reply.

An excellent source of information and advice is the Center for Disease Control, U.S. Department of Health, Education and Welfare, Public Health Service, Training Program, Community Services Training Section, Atlanta, Georgia, 30333.

Several well known manufacturers of foodservice industry equipment and supplies have produced films and film strips dealing with the sanitation routines to be followed in using their products. These are clear, practical examples of correct procedures, well worth viewing by both management and personnel. Ask your dish machine and detergent field technicians for information.

Write the Environmental Health Committee, Single Service Institute, 250 Park Avenue, New York, N.Y. 10017 for samples of their posters, handbook and other printed material—all are valuable aids for use in follow-up meetings.

The International Sanitary Supply Association, 5330 N. Elston Avenue, Chicago, Illinois 60630 has a variety of worthwhile sanitation material available to the customers of member firms.

Contact the National Restaurant Association, Educational Materials Center, 1530 N. Lake Shore Drive, Chicago, Illinois 60610 for a list of the food sanitation material developed by that organization. The films and sound-film strips make for interesting and informative viewing at follow-up meetings. Printed materials are also available from the NRA.

The National Sanitation Foundation, P.O. Box 1468, Ann Arbor, Michigan 48106 has published *A Reference Manual of Food Service Sanitation Educational and Training Materials* that any owner or manager would find very helpful. It is available for a modest purchase price.

The above listings are not intended as coverage of every source of visual aids for use in food sanitation training for personnel. Their purpose is to show foodservice management some of the choices from which to select attractive, interesting and worthwhile materials for the continuing sanitation training of foodservice workers.

The first follow-up meeting is an excellent time for review of accomplishments and failures during the period since completion of the training course. But such review must not be dull or pedantic or personnel will resent suggested corrections and become bored with the proceedings.

Preparations for Meeting No. 1

Let us assume the first meeting will be held four or five weeks after conclusion of the training course. During this time, personnel have been guided and corrected by management. Efforts have been made to change incorrect habits and procedures into required sanitary methods while the memory of discussed class material is still fresh.

During this interval, all those in positions of supervisory authority should make daily lists of personnel's sanitation errors, provide management with this information and any other facts relevant to the progress of the sanitation program.

Such lists will probably include many incorrect hands habits—putting fingers in the mouth, scratching the head, picking a pimple, using hand to dispense ice, rolls, butter; wiping hands on apron or service towel and later using same towel to wipe table, counter, work area; sampling food with hands, or spoon which is replaced in food without washing; smoking on the job; failure to wash hands after using the toilet; coughing or sneezing into food; other errors—leaving food exposed to dust, dirt, coughs, sneezes; failure to promptly refrigerate food; failure to properly clean equipment after use; faulty dishwashing procedures; faulty service procedures—mishandling of silverware, glasses, dishes; sweeping while food is exposed, etc.

Recurrence of these incidents should not discourage management. Long standing habits are hard to break. Only constant correction and discussion will change them.

About a week before the meeting, the meeting leader will list the most common errors. As many of these will be discussed as the planned time schedule will allow.

Decide that the first meeting will last approximately 45 minutes.

If the training course has been *Sanitation for Food Service Workers* and the slides are still available, the meeting leader can select slides illustrating these most common errors. He will review the lesson material pertaining to these slides until he knows exactly why the incorrect procedures are unsanitary and what unfortunate consequences can result if they are not corrected.

Remember, there will not be time for hesitation or lengthy discussions. Leader must be well prepared to stress key facts. The meeting must keep moving.

If slides are to be used in review, place them in order for showing, go over commentary, plan a few minutes for employee comment, time this part of the program to last *15 minutes.*

Timing can be assured if meeting leader practices in advance. Read error, allow time for personnel to correct orally, show pertinent slide, allow time for a minute's additional comment.

If some material must be omitted, plan to use it later, or—if urgent—let supervisors discuss it during working hours.

Alternative Suggestion

If slides are not available, plan for on-the-spot demonstrations. Select a few key personnel who can carry out the roles of careless individuals. Such employee participation is enjoyed by all.

Arrange a table set-up or any other props necessary. As employee acts out mistake, leader asks for personnel comment. What is wrong? What is correct way? Insist on rapid replies, move quickly from one employee to the next. If there is any doubt, or if incorrect answers are given, have acting employee demonstrate correct method and explain as she does.

Rehearse and time with selected employees, pledging them not to reveal meeting material before the program. Employees are usually happy to participate and will keep their part of it a surprise until meeting time.

Sometimes management prefers to combine employee demonstrations and slide showing using this approach: incorrect way demonstrated by employee; short question and answer period; correct way shown on screen; quick summary. If this method is to be followed, allow an extra 5 minutes for this part of the program.

Under the guidance of an enthusiastic, well grounded meeting leader, any of these methods or a combination of them may be used to create an atmosphere of interest, excitement, desire to respond and—as the most desired aftermath—a remembrance of the facts covered.

Note: The enthusiastic, well grounded meeting leader may be trained in any establishment. He or she must be an individual who believes in the importance of the program, enjoys talking to people, is willing to study the background material thoroughly, will experiment beforehand with timing, demonstrations, basic meeting setups such as equipment, placement, lighting, checking availability of extension cords, extra projector bulbs and all the other prosaic background necessities which comprise the holding of a smooth, successful meeting.

Plan to spend the next 15 minutes reviewing the section of the original training course, as indicated by errors, that is most, in need of further study.

If, for example, most errors fall in the area of personal hygiene, use personal hygiene slides, discuss material briefly, cover outstanding facts. Have employees participate here, ask them to take turns explaining slides, have them bring up any problems or material they do not thoroughly understand. If first part of meeting has been carried on as indicated above, employees will now be ready to make comments freely.

The Personal Hygiene group lends itself well to employee participation inasmuch as personal habits are involved. Spirited discussions can be held—are sanitary routines designed to speed good service? What is the effect of trained personnel on customers, the establishment as a whole? Why is the quality of food improved? What about flavor? The possibilities are endless.

If Personal Hygiene has been covered during discussion of sanitation errors in the first part of the program, use another section such as Food Protection, etc.

The meeting may conclude with the introduction of a contest or award system. Explanation of the contest should not require more than 5 minutes. It may run until the second meeting or may continue over several months. It may consist of points awarded for strict adherence to sanitary procedures, absence of checks for sanitation errors, credits for sanitary service; sanitation and courteous service may be combined. In other words, the contest may be set up in any way desired by management, may vary in different departments, just so interest is stimulated in the sanitation program and kept at high pitch by daily reminder of the contest.

Prizes or awards should always be worthy of the effort put forth by personnel.

Agenda for Meeting No. 1

1. Welcome and introductory remarks by meeting leader. (5 minutes or less)
2. Review of slides, based on sanitation errors. (15 minutes)

   or

   Review demonstrations illustrating sanitation errors (15 minutes)

   or

   Combination of slides and demonstrations. (20 minutes)

3. Review of section of training course, questions, comments, discussion. (15 minutes)
4. Explanation of contest. (5 minutes)
5. Concluding remarks. (5 minutes or less)

Time limits have been made flexible enough to allow for the few delays almost inevitable in a first meeting. The goal is to interest personnel but not weary them; to remind and correct, but to make such corrections palatable. Above all, to leave personnel with desire to return for a second meeting.

FOLLOW-UP MEETING NO. 2

General

These outlines are suggested guides, designed to stimulate management's thinking and assist in the development of meeting ideas which will help each particular group of personnel. Every establishment has its own problems. Work out as many of these problems as possible during your follow-up meetings.

Preparations for Meeting

1. Management will probably decide to hold the second follow-up meeting about a month or five weeks after the first session.

2. Plan to make this meeting entirely different from the first one. Arrange a showing of a moving picture or film-strip which will supplement the discussion topic. Make one group of employees responsible for the program. See to it that this group has enough preparation and rehearsal time to give a smooth interesting demonstration.

Since the work of the dishroom crew may be least appreciated by other personnel, why not build this program around the dishwashing department. Let the meeting adjourn to the dishroom where the crew will demonstrate proper dishwashing techniques from scrapping to after-sanitizing handling and storage.

As a crew member calls out his name and particular duty, a supervisor will point out how the correct sanitation procedures practiced at each station will result in clean, sanitized dishware and silver.

If time permits, have several of the watching personnel substitute at the machine for a few minutes each. Let them realize that alertness and dexterity are as necessary here as in any other department of the establishment. Be sure to call on a few women employees. Women have shown ability and speed in handling the duties of a modern dishroom.

In conclusion, point out that all the time, effort and care they have just seen will be wasted unless the remaining personnel will follow through with sanitary procedures when handling glasses, dishes and silverware.

3. How is your contest progressing? On the basis of your findings, plan for a new contest or a continuance of the current one. Have prizes for the first contest ready for the second meeting.

4. Be sure you are ready for this meeting. Study your posters, run the film through to be sure it is in perfect condition, have projector and screen set up in proper place, know what salient facts you are going to emphasize during the dishwashing demonstration.

The Meeting

1. The manager should open this second meeting. He will discuss briefly the previous month's operation from a sanitation standpoint. What progress has been made? How have customers reacted to the personnel's training in sanitary service? Has the flavor of the food been improved? Has it not been demonstrated that the sanitary method is actually more efficient and easier in the long run?

Personnel should be complimented on their progress, thanked for their cooperation and asked for continued support. Emphasize team work. When one employee fails, the entire program can be jeopardized. (5 minutes)

2. The meeting leader now takes over. Does any department need help? Are there any problems to be discussed? Tie in the new posters with this discussion; show them to personnel; decide where they should be hung for greatest effectiveness. (10 minutes)

3. Demonstration of proper dishwashing and storage techniques by dish room crew. (10 minutes)

4. Showing of film (15 to 20 minutes)

5. Awarding of prizes and discussion of new contest. (5 minutes)

This meeting could run over the 45 minute length, depending on the timing of the film. However, personnel will enjoy watching the film and should find no fault with the program set-up. If any part of the suggested program must be omitted, expand on the other discussion.

FOLLOW-UP MEETING NO.3

To maintain personnel interest and broaden the scope of their knowledge, the third meeting should be addressed by your health department sanitarian or, if preferred, by the head of the health department's food sanitation department. Your personnel should learn exactly how important their efforts are to the sanitation rating of your establishment. They must learn that concern for the sanitation program is exhibited in a very concrete way by someone besides management.

Hold this meeting no longer than six weeks after meeting No. 2. Perhaps by this time you will have decided that regular monthly meetings are most effective.

Preparations for Meeting

1. Contact your health department, arrange a mutually convenient meeting date. Tell your sanitarian what the meeting is about and outline the points you would like him to cover in his talk. You will find him eager to cooperate.

Your sanitarian is probably a familiar figure to your personnel and should conduct his talk on an informal, chatty basis. In non-technical terms he should discuss the necessity for food and drink ordinances—protection of community and personnel against food poisonings, communicable diseases, etc. He can then stress the tremendous importance of having management and personnel trained in sanitary procedures. He can point out a few requirements in the ordinance where personnel's training and knowledge will help make compliance so reasonable and understandable it will become much less difficult of accomplishment.

An explanation of the rating form and routine inspection procedure will follow, using the establishment's last rating sheet as an example. What items include personnel cooperation? Where did personnel failures cause checks for non-compliance? Where will personnel's correct procedure result in higher score for the establishment?

Unless self-inspection is already a part of the operation, the talk could well conclude with an explanation of this modern procedure and the suggestion that a self-inspection program be undertaken. Once a month management will make its own inspection on a regular health department rating sheet, using health department ratings. Following this, management will make its own corrections, thus easing the load on an invariably understaffed official agency while keeping management closely in touch with all phases of the sanitation program.

The once-a-month management tour for the self-inspection program should be one of many visits to various departments in the establishment. Eternal vigilance on the part of top management is the best guarantee for a successful food sanitation program in the foodservice operation.

The sanitarian must be sure to thank personnel for past cooperation and progress in maintaining sanitary standards.

2. Ask sanitarian to bring enough extra copies of your local food and drink ordinance for personnel to share in small groups.

3. Have a blackboard, chalk and eraser available. Your speaker will undoubtedly want to use them.

4. Be sure your most recent inspection report is available for reference.

5. Have contest awards ready. Decide whether to discontinue contests at this point or carry on for another month, perhaps picking up again later if indicated by personnel response.

The Meeting

1. Introduction of speaker by manager. (2 or 3 minutes)

2. Talk by sanitarian. (15 or 20 minutes)

3. Question and answer period. Led by supervisor, personnel should be encouraged to ask questions about the ordinance,

rating sheet items, scoring methods, self-inspection, etc. (10 minutes)

4. Manager thanks speaker for his talk, adjourns meeting. (5 minutes)

FOLLOW-UP MEETING NO.4

Subject–Hands

Preparations for Meeting

This meeting is designed to demonstrate just how well personnel have learned to communicate their sanitation knowledge to others. There need be no hesitation about attempting this meeting. It can be made enjoyable and both management and personnel can learn much from it.

1. You are going to have as many employees as possible explain briefly how their hands habits are involved with the wholesomeness of food, the prevention or cause of food poisonings and communicable diesases, or the creation of attractive foodservice. Different answers will be forthcoming from each department.

To spark enthusiasm and allow time for personnel to ready their answers, announce the meeting topic about a week in advance. Enlist the assistance of key personnel and department heads to help personnel prepare answers to questions on hand habits. If desired, a small prize could be awarded the department which, in the opinion of the entire group, makes the best showing. At conclusion of discussion, employees could indicate, by applause, which department has most impressed them.

2. If slides are available, select about a dozen to illustrate various uses of hands. Otherwise, use some of your posters or other new material for this purpose. Let employees identify reasons for showing each slide, connecting each one with his or her own duties.

3. Plan to have a few department heads–No. 1 chef, head storeroom man, No. 1 dishwasher, etc.–tell where his department has shown improvement in sanitation. If there are difficulties, where would they like help? Again, brief these individuals early so they can think about what they will say.

4. If self-inspection has been inaugurated, ask a few employees to tell how they contributed towards a good score.

The Meeting Agenda

1. Meeting leader announces topic for the day, explains that he will ask different employees about hands habits. Emphasize that answers are to be short and to the point. (2 or 3 minutes)

2. Question and answer period. (15 or 20 minutes)

3. Showing of slides or posters and discussion. (10 minutes)

4. Comments by department heads. Comments on self-inspection. (10 minutes)

5. Closing remarks. (2 or 3 minutes)

FOLLOW-UP MEETING NO.5

Subject–All about certain products and their correct usage.

This meeting will introduce a different approach to sanitation and, with a little careful preparation, can be made one of the most worthwhile of the series.

At first glance, it would seem to be intended for maintenance personnel only. This is not the case. All types of personnel have occasion to use some kind of cleaning material from time to time. In addition, personnel–particularly women–are quick to grasp how correct establishment usage can be profitably applied to home methods.

It is also well for all employees to learn how much knowledgeable detail must be assimilated and what a variety of products must be purchased by management to maintain the establishment in a sanitary condition.

Preparations for Meeting

1. Either of two approaches may be used to set up this meeting.

(a) If your main jobber or one of your suppliers is a member of the International Sanitary Supply Association, turn the meeting over to him and ask him to use ISSA material. This organization has excellent meeting scripts, posters, slide films and moving pictures on floor, wall, and window cleaning, dusting, etc., all of which are available to member firms.

(b) Contact your principal jobber or one or two product representatives and ask them to give short talks (about 10 minutes each). Products covered should include detergents, floor cleaners and how to use them, window and wall cleaning, dusting and polishing various metals, furniture and equipment.

Emphasize that these are *not* "selling" talks, inasmuch as you are already using the products. They are to be purely informatory and instructive. For example, why certain products are used for specific purposes while others are not permitted; how to sweep, mop, scrub, wash walls, clean windows, polish metals, dust–all in the most efficient manner.

It may surprise management to discover how many manufacturers of products and equipment designed for use in the food industry have prepared this type of material to ensure their products being used correctly to the best sanitation advantage. It may also surprise you to learn how little actual use is made of this material and how greatly your own personnel can benefit from a discussion and demonstration of correct product usage.

2. Insist on demonstrations of the proper use of products mentioned during the talks. Offer the services of your own employees for training as demonstrators.

3. Make a list of all items required for the meeting and check them off as they are assembled.

Meeting Agenda

1. Meeting leader's introduction of speaker or speakers and explanation of meeting topic. (2 or 3 minutes)
2. Talks and demonstrations of detergents, floor products and methods of floor cleaning, window and wall washing, dusting. If film is shown, about half of explanation and demonstration time will be allotted to this. (35 to 40 minutes)
3. Questions and comments by meeting leader and employees. (10 minutes)
4. Meeting leader thanks speakers, adjourns meeting. (2 or 3 minutes)

FOLLOW-UP MEETING NO. 6

Subject–More about products and their correct usage.

Do not allow too much time to elapse between this meeting and the previous one. Personnel will more easily retain an over all grasp of product variety and usage.

The purposes of meetings 5 and 6 are:

1. To familiarize personnel with the sanitation tools and products provided by management to help them maintain the premises in a sanitary condition.
2. To review or, in some cases, instruct in correct product usage so the desired goal will be achieved.
3. To secure interest and cooperation through frank discussions of the products involved.

The cost to management when products are wasted or misused through personnel misunderstanding or carelessness should be included.

The meeting leader may stress these points in his opening remarks.

Preparations for Meeting

Contact for this meeting, product representatives who can discuss rodent and insect control products; the tools, products and methods necessary to maintain washrooms in immaculate condition; and, finally, someone who can give a short, forceful talk on the maintenance of garbage cans and garbage rooms.

If rodent and insect control is a problem and extra material is thought necessary, the Center for Disease Control, Atlanta, Ga., can provide very interesting visual aids, including films. In this case, you may wish to dispense with a speaker on this subject and show a film instead.

The International Sanitary Supply Association, 5330 N. Elston Ave., Chicago, Ill. 60630 has excellent material on washroom sanitation which your speaker would find invaluable.

The Soap and Detergent Association (Industrial Division), 295 Madison Avenue, New York, N.Y. 10017 issues periodical portfolios on "Tips on Cleaning." These are interesting, well illustrated collections of facts and information about every type of cleaner and how to use it. They would appeal to both management and personnel.

The time devoted to rodent and insect control should be sure to include a description of the establishment's pesticides, their correct usage and possible misuse, such as contamination of food, dying flies falling into food, etc.

Washroom sanitation has been included because this is an area in which all personnel are directly or indirectly involved. Have your speaker be prepared for a thorough coverage here, from odor control to wall and equipment cleaning.

When personnel understand the time and effort required to maintain garbage cans and storage areas in a clean, sanitary condition, it will be easier to ensure their cooperation in the care of cans and lids so that denting of cans, loss of lids and other abuses can be corrected. Garbage can sanitation is also an important factor in rodent and insect control. A description of the cleaning of cans and garbage room by the maintenance man involved (use of steam hose, etc.), plus the mention of the presence of various types of silverware and other establishment items in the garbage will impress personnel with the necessity for care.

If a film is used, the meeting agenda must be adjusted accordingly. These meetings can be very flexible, no long scholarly discussions are required, but short, interest filled facts are necessary, with every sentence meaning something and adding to the discussion.

Meeting Agenda

1. Opening remarks by meeting leader, introduction of speakers. (5 minutes)
2. Rodent and insect control products and usage. (10 minutes)
3. Washroom sanitation. (10-15minutes)
4. Garbage can and garbage room sanitation (8 to 10 minutes)
5. Discussion period, questions and answers. (5 to 10 minutes)
6. Meeting leader thanks speakers, adjournment. (3 minutes)

FOLLOW-UP MEETING NO. 7

Subject—Teamwork.

Discussion—During their initial training and previous follow-up meetings, your personnel have undoubtedly heard much about the necessity for teamwork. Now is the time to demonstrate this necessity in a concrete manner.

Why teamwork? Because both following, or neglecting to follow, sanitary procedures can have a direct effect on the health of customer and personnel. And a breakdown in the sanitation program may occur at any point in the establishment's chain of operation, whenever a single employee becomes careless. Therefore, 100 percent cooperativeness can be considered more critical here than in any other business enterprise—because health is involved, and involvement depends almost entirely on the actions of the personnel.

In foodservice, personnel teamwork ensures a smooth over-all performance, better service, more desirable working situations and a knowledgeable attitude about the other fellow's job—simply because all are working together.

The management-personnel team factor is also an important consideration. Management cooperates by providing sanitary physical surroundings, equipment and tools, and by training personnel in sanitary procedures. Personnel cooperate by maintaining, surroundings, equipment and tools in the required sanitary condition and by observing the sanitation rules and procedures learned during the training sessions.

Preparations for Meeting

Select two popular items from your regular menu, preferably one which is cooked and one served uncooked. You may choose anything from chopped beef through prime rib, ham, fowl, seafood, etc.

Once you have decided, follow this item through your establishment from the time of its delivery until it is served to the customer.

Who checks in this food item? Who is responsible for its refrigeration? Is it kept in the freezer or stored elsewhere in the icebox? Who checks refrigeration temperatures and keeps the refrigeration clean?

Next comes preparation. If the food is frozen, who is responsible for thawing it properly? Who cooks it? Who keeps the cooking equipment and utensils in a sanitary condition?

Is the food served immediately after cooking or held on the range, steam table or bain marie? Who checks for correct holding temperatures?

Who is involved in placing portions on

the service plate? And who is responsible for providing a sanitary plate for the food?

Who sets up the table and who serves the food?

List all personnel concerned. Now you have your team of personnel involved in bringing this single food item to the customer in a sanitary condition.

Follow this same procedure with your uncooked item, which could be something as simple as a head of lettuce. First steps as to delivery and storage will be the same as for Example 1. Then you will stress responsibility for thorough washing (and clean hands), proper method of preparation to prevent contamination, correct preservice temperatures, then serving, etc. as with the cooked item.

If the "head of lettuce" idea is not challenging enough, try the ingredients of any raw salad; or any other raw menu item of your choice.

Your second list will constitute another personnel sanitation team. Certain employees will, of course, be on both lists, a fact which you can use to emphasize their importance in the over-all sanitation program.

Prepare a 2 or 3 minute opening talk covering the points mentioned above under Discussion.

Have a blackboard on which you can write your lists of personnel.

Make every effort to secure a full attendance for this meeting, since every department is involved.

Agenda of Meeting

1. Meeting leader explains topic—teamwork. He outlines how team work will be demonstrated by using two menu items as illustrations. (5 minutes)

2. Beginning with cooked item, meeting leader lists the various steps taken to bring item from storage to dining room, asking for suggestions from personnel as he writes. (8 minutes)

3. As he mentions each step, employee concerned will either rise, or leave seat and come to front of room. To avoid more than one waiter or waitress getting up, leader can specify one particular serving station.

4. When sanitation team has been assembled, ask respective team members to mention various ways in which the menu item could have been contaminated had they not observed sanitary procedures. Ask remaining personnel to listen carefully and supply any information overlooked by team members. (10 to 15 minutes)

5. Repeat for menu item no. 2—the uncooked food. Since personnel know what to expect this second time, listing and procedure enumeration should move along more rapidly. (10 minutes)

6. Conclude with further group discussion of teamwork. Ask personnel for more illustrations. (Management, for example, is part of the team here by its purchase of quality foods.) Compliment personnel on team work exhibited thus far and ask for continued cooperation. (5 minutes)

7. Adjournment.

FOLLOW-UP MEETING NO. 8

Subject—A discussion of communicable diseases and food poisoning.

Discussion—Bring in an outside speaker, preferably the Health Officer or one of his assistants qualified to discuss these two topics. The entire meeting will be devoted to this talk.

Preparations for Meeting

Contact the health officer and discuss this meeting. If the health officer is unable to attend, he will probably suggest an assistant as substitute, or a local physician.

As with the meeting addressed by your sanitarian, you should be specific about what you want. For example:

Certain communicable diseases are definitely capable of transmission through contaminated food, insanitary utensils and/or the careless habits of unthinking personnel. This can be extremely costly to management, personnel and customers.

The cost of the common cold alone is estimated to be $5 billion a year in lost wages and production and in medical expenses. Workers in foodservice contract their share of respiratory infections. Some of these remain at home when feeling sick, many others do not. These latter may be responsible for infecting fellow workers or customers who may become seriously ill.

Thus, an interesting discussion of the common cold and the other relevant communicable diseases will refresh the memories of personnel on the points covered in the communicable disease training session. An address delivered from the medical man's viewpoint will impress and carry great weight with the employees.

As for food poisonings, still the bugaboo of the foodservice industry, a vivid description of a few outbreaks, their causes and consequences will assure a most attentive audience—always provided the speaker does not indulge in technicalities and terms beyond the understanding of his listeners.

These are matters to be decided upon in advance with the speaker.

Agenda

1. Meeting leader introduces speaker. (3 minutes)
2. A doctor discusses food poisonings and communicable diseases associated with foodservice. (30 minutes)
3. Question and answer period (led by meeting leader). (10 minutes)
4. Adjournment.

FOLLOW-UP MEETING NO. 9

Subject—Self-Inspection

(a) An explanation of self-inspection.
(b) Personnel participate in a self-inspection tour of the establishment.

Preparations for Meeting

1. Secure enough health department inspection sheets to permit sharing among groups of 4 to 6 employees. Have pencils available for each group. If possible, have flashlights ready for use in areas where your regular inspector would need them.
2. Depending upon number of personnel, plan to divide employees into working groups for self inspection tour. If total number does not exceed 10 or 12, this division will not be absolutely necessary.
3. Review health department inspection sheet items explained to personnel during meetings No. 3 and 4. Basing remarks on present status of self-inspection program and degree of emphasis placed on employee attention to rating sheet items, plan a short introductory talk. Remind employees what the sanitarian looks for during an official inspection, then tell them they are about to conduct a similar inspection of their own.
4. To avoid congestion in any one area, plan route for inspection teams, again depending on number of groups involved. The purpose of this tour is serious—to learn just how much personnel know about official sanitation requirements and how quickly errors can be spotted as they tour the premises.

In some instances where this same theme meeting has been held in large operations, management has left some employees on the job.

The employees at work have been instructed to make deliberate personal sanitation errors during the team inspections, as well as to set up incorrect situations such as walk-in box door left open, food exposed, utensils and equipment left unwashed after use, food spilled on the floor, cold water in pot washing sink, etc. This seems an excellent idea if it does not conflict with a time of actual operation.

5. Select supervisors or key employees as group leaders.

Agenda

1. Introduction, with explanation of meeting. (5 minutes)
2. Division into groups, with directions to each group as to route to follow through establishment. Answer any questions. (5 to 10 minutes)
3. Teams inspect establishment, mark sheets. Meeting leader moves from team to team, checks his own inspection sheet. (20 minutes)
4. Teams reassemble, meeting leader checks sheets, compares them for accuracy against his own. Discussion period follows. Why were certain items marked on some sheets and not on others? Has the inspection resulted in a good score for the establishment? If not, what corrections can be made immediately? For how many errors are personnel responsible? Do employees now have a clear understanding of rating sheet and what is expected of them? (10 minutes)
5. Adjournment.

FOLLOW-UP MEETING NO.10

Subject—Sanitation-
1. Money saver
2. Time saver
3. Aid to safety
4. Quality improver of food and service
5. Creator of desirable public image for establishment (management) and personnel

Preparations for Meeting

1. Have the blackboard ready for this meeting.

2. Use posters around the room illustrating the various points listed above. (See outline of meeting No. 2 for visual aid sources.)

3. Under the 5 headings given above, list pertinent items. On meeting day, this will be done on the blackboard. Plan how to encourage personnel to contribute to the lists. At this point, they should have a sufficient grasp of the subject to offer valuable suggestions.

Examples of points to be included and developed by leader:

1. Sanitation as a money saver.
   (a) Equipment and utensils properly used and maintained last longer.
   (b) Foods which have been stored, refrigerated, prepared and served in accordance with sanitation rules prevent waste, enhance quality, improve in appearance, increase customer count.
   (c) Danger of food poisonings, with consequent cost to management, reduced to a minimum.
2. Sanitation as a time saver.
   (a) The sanitary way is the *right* way.
   (b) Experience has shown the right way to be the quickest way.
   Consistent practice in sanitary methods has proved this.
3. Sanitation as aid to safety.
   (a) Proper mopping times and methods minimize danger from falls on wet, slippery floors.
   (b) When floors are kept clean, employees cannot slip on dropped and spilled pieces of food.
   (c) Correct methods of handling sharp knives, dishes, glassware, silverware will reduce danger of cuts and infections.
   (d) Strict adherence to sanitary procedures will also prevent transmission of illnesses commonly associated with food service.
4. Sanitation improves quality of food and service.
   (a) Enlarge on item "b" under "money saver." Sanitary methods prevent serving of "sick" (contaminated) food. Safe food, clean equipment and utensils, ensure true, appetizing food flavors.
   (b) Sanitary service is *correct* service. Personnel who know they are serving correctly develop the confidence and poise inherent in quality service.
5. Sanitation as a creator of desirable public image.
   (a) A sanitary establishment never fails to create a favorable atmosphere.
   (b) Professional appearance and methods of trained foodservice worker are always noticed by customer. Adept and correct hand habits used when serving food create a most favorable reaction.

The above are suggestions from what could become a much longer listing. Points and situations particularly applicable to your own operation should be enumerated, with attention also given to any other ideas contributed by personnel. Make your list as long as time permits.

Agenda

Introduction—Explain that the purpose of the meeting is to illustrate the concrete value of the sanitation program and obtain personnel's honest opinions and suggestions. Write the subject and sub-headings on the blackboard. (10 minutes)

Discussion—Taking each sub-heading in turn, write points to be included on the blackboard, pausing for discussion and comment on employees' suggestions. There may, for instance, be objections to the statement that the sanitary way is the quickest way. If so, it will be necessary to point out that any insanitary "quick" methods are only sloppy substitutes for the orderly, systematic and sanitary work habits of the professional foodservice employee. (20 minutes)

Adjournment.

FOLLOW-UP MEETING NO. 11

Subject—Background and development of the establishment (or chain), with emphasis on the development of sanitation over the years.

Preparations for Meeting

Most employees are naturally curious about the history and growth of the establishment or chain for which they work. Personnel should be furnished this information; it will give them a feeling of importance to know they are an integral part of a successful business enterprise. Knowing the background of their establishment will provide greater understanding of their own operation and stimulate them to contribute their best efforts to the further success of the foodservice. The encouragement of such an attitude is particularly necessary in foodservice, where too many employees regard their work as "just another job."

It would be most effective if the manager, or—in the case of a chain operation—a visiting company official could give this talk to the employees.

Outline the beginnings of the business and trace its expansion to the present time, giving due credit to personnel for assisting in this development. If any photographs or news releases concerning the opening or later events are available, they will be greatly enjoyed during this meeting.

Along with the growth of the business will come a description of the advancements in sanitation. If the founding of the business occurred 35 or 40 years ago, this will be extremely interesting to personnel, particularly if the speaker is frank in discussing the first methods and problems. The sanitation methods, probably rudimentary, practiced in the beginning *are not to be regarded as deficiencies on the part of management* because this was not the case.

Official agency knowledge of sanitation requirements in the safe preparation of food and drink, as well as the necessity for today's dishwashing and sanitizing methods and temperatures, construction of utensils and equipment, etc. has kept pace with the modern sanitation requirements made necessary by developments in techniques, equipment and products.

Management should point out, therefore, how—over a period of years—this increasing sanitation knowledge and its application to foodservice procedures actually enhanced the quality of the operation's food and service and thus contributed materially to the growth of the business. New equipment, different methods, even extensive building alterations that may all have been required at one time or another were undoubtedly very expensive and may even have met with temporary opposition by management, who might have failed at first to understand the necessity for these innovations. But in today's competitive foodservice market, the advantage of being well equipped and well trained sanitation-wise is obvious, with the results evident every day in the cash register.

Employees are always interested in participating in discussions regarding establishment policy, especially where they are concerned. Meeting leader will already know management's decision regarding a second training school and can guide opinions and comments in the desired direction.

If there has been a sizable personnel turnover, such a course is definitely indicated. Supervisory help cannot take the place of the complete series of lectures, visual aids and sanitation information gained from a regular training course.

If the personnel roster remains practically the same, employees can be told that certain course topics—where personnel are weak—will be selected for review. This second time, varying methods of presentation will be used. Old time employees will be expected to assist by operating slide projector, assisting in slide showings, putting on demonstrations, even leading some of the discussions.

When employees learn that a fresh approach will be used, new facts added, etc. they will look forward to further training sessions. Ask them to list anything they wish to review or anything they do not fully understand. Thus they will have a concrete part in choosing some of the material to be covered.

This meeting can be conducted satisfactorily within 40 minutes.

Closing Comments

1. Discussion in Meeting No. 11 is based on the assumption that management has conducted its own training course—*Sanitation for Foodservice Workers,* for example. In many areas, health departments require newly employed personnel to attend health department training sessions unless they already have local attendance record cards.

Under this set-up, management can still hold its own review classes whenever necessary by obtaining a copy of Sanitation for Foodservice Workers or other selected material.

2. The meeting outlines suggested here may be held between regular review sessions to vary the format and secure more interest and responsiveness than a steady diet of straight review material could possibly achieve. The order of the meetings can also be changed in any way desired (with the exception of No. 12).

The general coverage of management's talk has been discussed in the preceding paragraphs. No doubt personnel will expect it to be entertaining, humorous and filled with anecdotes of past foodservice experiences.

It should be all of this and more. Concrete facts on the esthetic, health promoting and financial advantages of maintaining a sanitary foodservice operation must be provided.

It is true that the inspecting sanitarian corrects personnel during each visit; but advantages to be gained by foodservice workers, when they acquire modern sanitation expertise, will be considered much more thoughtfully when presented by top management. Therefore, the speaker should emphasize how important the development and practice of food sanitation have been to the success and continuation of the foodservice industry—now one of the nation's foremost business enterprises.

In closing, the speaker should ask for questions. If led by the meeting leader and a few key personnel, questions about past events in the establishment's history are usually numerous.

Agenda

1. Introduction of speaker and explanation of subject by meeting leader (3 minutes)
2. Talk by management (20 minutes)
3. Discussion, question and answer period (10 minutes)
4. Adjournment.

FOLLOW-UP MEETING NO. 12

Subject—A look back. A look ahead.

Discussion—Prior to this meeting, management should consider whether a second sanitation training course for personnel would be advisable in the near future. A decision could well be based on the number of new personnel who have not had this training; and on the sanitation progress of old employees. If a decision has been reached, the meeting leader should be notified before this meeting.

Preparations—The look back

Since this is the 12th meeting after completion of the training course, a brief but different type of review is indicated.

1. Plan to request cooperation of a few key personnel to assist with suggestions. (See next page).

2. Plan a review talk around personnel accomplishments and the advantages of a sanitation program, always emphasizing a positive approach. Leave obvious personnel failures, if any, till the second part of the program.

3. What has the sanitation program done for personnel?
   (a) Provided more attractive working situations.
      (i) Consistently clean surroundings are more pleasant.
      (ii) Properly prepared food is safe to eat. (Personnel's food is as important as that served to customers.
      (iii) Correct hands habits and personal hygiene observances materially decrease chances of personnel infections, absenteeism through illness.
      (iv) Sanitary maintenance methods decrease chances of personnel accidents on the job.
   (b) Sanitary practices increase efficiency.
      (i) Continuous concentration on correct methods inevitably leads to increase in efficiency.
      (ii) Facts learned in follow-up meetings, sanitation team cooperation, knowledge gained of requirements in other departments all lead to over-all personnel efficiency.
   (c) Necessity for change of working habits results in increased alertness, provides valuable training in self-discipline.
   (d) Active participation in sanitation program increases interest in doing job well, prepares employee for advancement.

Above points may be briefed for listing on a blackboard. Personnel participation in contributing to list of suggestions will be sparked if key personnel have been alerted to supply some of the desired answers. Personnel will undoubtedly have suggestions of their own.

Preparations–The look ahead

1. Here is where deficiencies can be listed.
   (a) In what areas can we improve even more?
      (i) Handwashing?
      (ii) Personal hygiene habits?
      (iii) Dishwashing?
      (iv) Refrigeration practices?
      (v) Food preparation?

The meeting leader knows where weaknesses exist. He can suggest various departments without embarrassing any single employee by mentioning names. Personnel will understand the references.

   (b) What shall we do to strengthen our weak areas?
      (i) How much of a personnel turnover has occurred?
      (ii) Are new personnel being sufficiently helped and taught sanitary methods by supervisors and old em- employees?
      (iii) Is this type of training enough or should we hold some regular training sessions for new personnel? Such meetings could also act as refresher courses for old employees in need of help.

BACTERIOLOGY QUIZ ANSWERS

1. Both are alive. Both require food and moisture. Both give off wastes.
2. Size. Reproduction. Method of eating.
3. Cocci, Bacilli, Spirilla.
4. (c) is correct.
5. Virus, protozoa, molds.
6. b, c, are correct.
7. No.
8. (c)
9. Toxins.
10. Some cause disease.

COMMUNICABLE DISEASE QUIZ ANSWERS

1. (1) By inhaling droplets from coughing or sneezing.
   (2) By eating food contaminated by coughs, sneezes, unclean hands.
   (3) By mouth contact with contaminated cups, glasses, eating utensils, etc.
2. (b)
3. (a) False. (b) False. (c) True.
4. Communicable
5. (1) Hands. (2) Handkerchief. (3) Disposable tissue.
6. (d) is correct.
7. carrier
8. The hands
9. (1) After restroom visits. (2) After coughing, sneezing, smoking.
   (3) After handling of customer soiled utensils, or soil of any kind.
10. "catching" or "contagious."

FOOD-BORNE DISEASES QUIZ ANSWERS

1. (a) False. (b) True. (c) True.
2. (a) Staphlococcus, Clostridium perfringens. (b) botulism
3. (a), (c) are correct statements.
4. Bacteria (Parasites also correct but not so common.)
5. Any three of custards, seafoods, meats, milk and milk products, eggs, poultry and poultry dressings, salad dressings, shellfish, raw salads and raw vegetables, cream-filled pastries.
6. (a) False. (b) True. (c) True.
7. (b), (d) are correct statements.
8. 45°F. and 140°F. . . . . .danger or incubation zone.
9. time, center, shallow, freezer.
10. (a) False. (b) True.

FOOD PROTECTION QUIZ ANSWERS

1. (a) True. (b) False. (c) False.
2. (a) immediately discarded. (b) disease-causing.
3. (a) is correct statement.
4. 150°F.
5. (a) False. (b) True. (c) True. (d) False.
6. (a), (c) are correct statements.
7. (a) pressure-cooker.
   (b) removing moisture.
   (c) meats, fish.
8. No.
9. 169 million.
10. (a) True. (b) False. (c) False. (d) True.

PEST CONTROL QUIZ ANSWERS

1. (b), (d) are correct.
2. hairy, sticky substance, feet.
3. (c)
4. tuberculosis, typhoid fever, dysentery, salmonellosis.
5. walking over food or rubbing against it, regurgitation while eating, waste elimination.
6. No
7. (b)
8. dysentery, diarrhea, typhoid fever, food poisoning.
9. easily available food supply, suitable nesting place to breed their young.
10. personnel, foodservice.

MANUAL DISHWASHING QUIZ ANSWERS

1. three, washing, rinsing, sanitizing.
2. (a), (c) are correct statements.
3. (a) True. (b) False. (c) True. (d) False.
4. (a) cracked. (b) soak. (c) freshly deposited.
5. (b), (c) are correct.
6. (a) True. (b) False. (c) True.
7. (a), (b) are correct statements.
8. (a) soaking. "To be washed with brightening or stain removing detergents" would also be correct answer. (b) bacterial.
9. (a) True. (b) False.
10. True.

DISHWASHING COMPOUND QUIZ ANSWERS

1. Detergents, personnel, equipment.
2. (a), (c), (d) are correct.
3. True.
4. True.
5. False.
6. True.
7. (a) particles, suspended, dishware, (b) water, wood ashes, grease.
8. (a) and (c) are correct.
9. True.
10. False.

DETERGENT DISPENSER QUIZ ANSWERS

1. Dishwashing compound, wash solution in wash tank.
2. Strength, wash solution.
3. (a), (c) are correct statements.
4. True.
5. False.
6. (1) controller, (2) solenoid valve, (3) electrode.
7. (1) Dispensing detergents directly from shipping container to dishmachine's wash tank.
   (2) Dispensing detergents from remote area to dishmachine's wash tank.
8. True.
9. False.
10. True.

MECHANICAL DISHWASHING QUIZ ANSWERS

1. (a) wash, rinse, sanitize.
   (b) Roll-type, door-type, conveyor-type, multiple tank conveyor type.
2. (a), (b), (c) are correct statements.
3. (a) True. (b) False. (c) True. (d) False.
4. (a) food soil. (b) 110°F. to 130°F. (c) 150°F., 160°F., 170°F.
   (d) 160°F. - 170°F. (e) 180°F.
5. (a) False. (b) True. (c) False. (d) True.
6. (a), (b) are correct statements.
7. (a) crowding or over-lapping.
   (b) inverted.
   (c) handles.
8. (a) False. (b) True. (c) False.
9. (a), (c) are correct.
10. (a) True. (b) True. (c) True.

SANITIZING QUIZ ANSWERS

1. (a) True. (b) False. (c) False.
2. alkaline, acid.
3. (a), (c) are correct statements.
4. No. 3 hot water, chemicals, one-half minute.
   at least one, chlorine, iodine.
5. (a) True. (b) False. (c) True.
6. sanitizing, detergent-germicide, disinfectant-sanitizer.
7. (a) True. (b) False. (c) True.
8. (a) dry, floor.
   (b) splash from mopping solutions, food spillage, dust, floor dirt.
   (c) inverted.
9. (a), (c) are correct statements.
10. (a) True. (b) False. (c) True. (d) False.

DISHROOM MAINTENANCE QUIZ ANSWERS

1. cleaning solution, sufficient force, direction.
2. (a) True. (b) True. (c) False. (d) True. (e) False.
3. (a), (c) are correct statements.
4. (a) often, hot water supply line.
   (b) discolored, discolorations.
5. (a) False, they should be kept clean.
   (b) True.
   (c) True.
   (d) True.
6. (b), (c) are correct statements.
7. well lighted, well ventilated.
8. (a) True. (b) True.
9. (a), (c) are correct statements.
10. (a) True. (b) True.

PERSONAL HYGIENE QUIZ ANSWERS

1. (1) Thorough physical check-up once a year, including dental and eye examinations.
   (2) Eat proper amounts of the right foods.
   (3) Get plenty of rest at night.
2. (1) Daily bath, frequent shampoos.
   (2) Clean teeth daily.
   (3) Use deodorant and mouthwash daily.
3. hands habits.
4. (c), (f) are correct statements.
5. (c), (d), (f) are correct statements.
6. (d) is correct statement.
7. (a), (b), (c) are correct statements.
8. (1) Lack of knowledge.
   (2) Carelessness.
   (3) Unwillingness to cooperate and do the job right.
9. Fill in answers:
   (a) underclothing
   (b) sensible shoes
   (c) hair net, cap
   (d) jewelry
   (e) make-up
   (f) uniform
   (g) hands, nails, nail polish
   (h) hose, runs, seams
   (i) chew gum
   (j) smoking
10. Yes
   (1) Practice of sanitary habits helps protect health of customers, fellow workers.
   (2) Good personal hygiene habits also protect food.
   (3) Correct habits do much to promote desired "image" of establishment as an enjoyable place in which to eat.

REFERENCES

The ABC's of Microwave Cooking. Reprint from Microwave Applications Newsletter, P.O. Box 241, Amherst, N. H. 03031.

Bryan, Frank L., Ph. D., M.P.H., and Kilpatrick, Edward G., B.B. *Clostridium Perfringens Related to Roast Beef Cooking, Storage and Contamination in a Fast Food Restaurant.* Reprinted by the U.S. Department of Health, Education and Welfare from the American Journal of Public Health, Vol. 61, No. 9, September, 1971.

Dack, G. M., Ph. D., M.D., *Food Poisoning.*

Rat Borne Disease Prevention and Control, Public Health Service Center for Disease Control, Atlanta, Georgia.

Stevens, Roger, et al. *The Microbial World.* Prentice-Hall, Englewood Cliffs, N.J.

176 2 by 2 inch slides, illustrating the topics discussed in SANITATION FOR FOODSERVICE WORKERS, are available for rental or purchase. Please write Mrs. Treva M. Richardson, 220 S. Logan Street, Mason City, Illinois 62664 for details on rental and purchase procedures.

A set of printed commentaries is included in the slide package. These suggest brief comments for each picture and serve to coordinate the slides with the textbook material.

Protecting customers from food contaminated by harmful bacteria is a responsibility shared by all foodservice employees. The health of the girl in the foreground is in danger if her cup has been contaminated during or after washing. Learning how to protect her is made easier for workers who learn and follow the principles outlined in this book.